P. A. Krupchitsky

Fundamental Research
with
Polarized Slow Neutrons

Translated by V. I. Kisin

With 42 Figures

Springer-Verlag Berlin Heidelberg New York
London Paris Tokyo

Pyotr A. Krupchitsky
25 Bolshaya Cheremushkinskaya St., 117259 Moscow, USSR

Translator:
Vitaly Isayevich Kisin
24 Varga St., Apt. 9, 117133 Moscow, USSR

ISBN 3-540-16996-2 Springer-Verlag Berlin Heidelberg New York
ISBN 0-387-16996-2 Springer-Verlag New York Berlin Heidelberg

Library of Congress Cataloging-in-Publication Data. Krupchitsky, P.A. Fundamental research with polarized slow neutrons. Translation of: Fundamental'nye issledovaniia s poliarizovannymi medlennymi neĭtronami. Bibliography: p. Includes index. 1. Thermal neutrons. 2. Polarized beams. I. Title. QC793.5.T422K7813 1987 539.7'213 86-20392

This work is subject to copyright. All rights are reserved, whether the whole or part of the material is concerned, specifically those of translation, reprinting, reuse of illustrations, broadcasting, reproduction by photocopying machine or similar means, and storage in data banks. Under § 54 of the German Copyright Law where copies are made for other than private use, a fee is payable to "Verwertungsgesellschaft Wort", Munich.

© Springer-Verlag Berlin Heidelberg 1987
Printed in Germany

The use of registered names, trademarks, etc. in this publication does not imply, even in the absence of a specific statement, that such names are exempt from the relevant protective laws and regulations and therefore free for general use.

Offset printing: Druckhaus Beltz, 6944 Hemsbach/Bergstr. Bookbinding: J. Schäffer OHG, 6718 Grünstadt.
2153/3150-543210

Preface

Nearly 20 years have elapsed since the publication of the monograph *Polarized Slow Neutrons* by Yu.G. Abov, A.D. Gulko, and P.A. Krupchitsky (Atomizdat, Moscow 1966). In the preface the authors emphasized that the advent of polarized beams of slow neutrons led to new techniques widely used in nuclear and solid-state physics. Now, two decades later, polarized slow neutrons are used even more extensively in fundamental research in nuclear physics. Suffice it to mention that the effect of parity violation in nuclear interactions, discovered by using polarized neutrons, was subjected in this period to a thorough, comprehensive scrutiny, receiving solid confirmation that led to the discovery of parity violation in the fission of nuclei and in neutron optics. Further progress was achieved in the search for the electric dipole moment of the neutron and in the study of the beta decay of the neutron. Besides, the nuclear precession of neutrons and the coherent interference of spin channels in the neutron capture by nuclei were discovered.

The descriptions of the relevant experiments are dispersed among numerous journals, preprints, and specialized reviews. This situation is only partially remedied by the recently published monograph *The Fundamental Properties of the Neutron* by Yu.A. Aleksandrov (Energoizdat, Moscow 1982). The need for a new book on fundamental research with polarized slow neutrons is thus quite obvious.

There are several aspects to the research involving polarized slow neutrons. First, it is concerned with the fundamental properties of the neutron as an elementary particle and with the fundamental interactions in which it participates. Second, polarized neutrons constitute a powerful tool for studying the structure of nuclei and the properties of nuclear forces and nuclear reactions. Consequently, polarized neutron research is a field in which the problems of elementary particle physics intertwine with those of nuclear physics.

It should be emphasized that this combination is far from formal, because it has been demonstrated that correctly selected nuclei constitute a laboratory in which small effects due to the weak interaction are amplified, thus becoming accessible to measurement. The present book is primarily devoted to these two aspects of research with polarized neutrons.

The third aspect, namely, the study of the condensed state of matter, is now evolving into an independent branch of physics and will not be discussed in this monograph.

The book would be incomplete if we failed to describe the new techniques, developed in recent years, of generating and analyzing polarized neutron beams. For this reason, Chap. 1 briefly summarizes the methods employed when working with polarized slow neutrons, also covering such novel devices as polarizing neutron guides.

The remainder of the book consists of two parts. Chapters 2–5 chiefly treat fundamental research not related to the violation of quantum-mechanical conservation laws. Chapter 6 then introduces the quantum-mechanical properties of symmetry. These properties are necessary for interpreting polarized neutron experiments in which conservation laws (for example, the conservation of spatial parity or invariance under time reversal) are violated (Chap. 7–12).

Chapter 7 describes experiments on the decay of polarized neutrons, which were important for the elaboration of the weak interaction theory and, in particular, of the $V-A$ version of the theory. Chapter 8 is devoted to spatial parity violation in nuclear interactions; this effect was manifested by the anisotropy of γ-radiation emitted as a result of the capture of polarized neutrons by nuclei. Proceeding from anisotropy in the distribution of fission fragments, further progress in this area of research led to the discovery of parity violation in the fission of nuclei (Chap. 11), the search for parity violation in reactions emitting α-particles or other light nuclei (Chap. 10), and, finally, the discovery of parity violation in neutron optics experiments (Chap. 12).

The book thus aims at isolating specific physical phenomena arising in experiments with polarized neutrons. This organizational principle makes it possible to give a consistent description of the basic method used to study these phenomena, review the relevant experiments with particular emphasis on the latest results, and summarize the data of reported measurements. It is not always possible to achieve the latter because the number of available results often runs into many hundreds, as, for example, in the case of the measured characteristics of nuclear levels (Chaps. 4 and 5).

An even more difficult task in a book of this type is the theoretical generalization of experimental results. This generalization would inevitably involve a great deal of experimental data obtained for nuclear reactions with either nonpolarized neutrons or, in some cases, no neutrons at all. Accordingly, where necessary, I refer the reader to review papers cited in the reference list.

My principal objective was to demonstrate the similarity of experimental approaches employed in fundamental research with polarized slow neutrons and, at the same time, to point out the difficulties that have to be overcome in order to analyze negligibly small effects in experiments measuring various radiation asymmetry coefficients. I have assumed that the reader is familiar with the basic concepts of neutron physics. The most detailed presentation of these concepts can be found, for example, in *Low-Energy Neutron Physics,* by I.I. Gurevich and L.V. Tarasov (North-Holland, Amsterdam 1968).

I hope that the book will prove to be of help not only to specialists in neutron physics, but also to all physicists who are interested in understanding the role and the specifics of fundamental research in general.

In conclusion, I wish to express my gratitude to Yu.G. Abov and A.D. Gulko for permission to use portions of material contained in our earlier book, to G.V. Danilyan for fruitful discussions on a number of points, and also to O.N. Ermakov, I.L. Karpikhin, and V.F. Perepelitsa, with whom I carried out some of the studies described in this book.

P.A. Krupchitsky

Table of Contents

1. Brief Summary of Experimental Techniques Required for Working with Polarized Slow Neutrons 1
  1.1 Generation and Analysis of Polarized Slow Neutron Beams 1
    1.1.1 Behavior of Neutron Spins in Magnetic Fields 1
    1.1.2 Schematic Diagram of a System for Generating and Analyzing Polarized Neutrons 3
    1.1.3 Measurement of Beam Polarization 4
  1.2 Experimental Methods for Studying Polarized Slow Neutron Beams 5
    1.2.1 Collimators, Filters, Detectors 5
    1.2.2 Adiabatic Spin Rotation 7
    1.2.3 Spin Flippers 8
  1.3 New Methods of Generating Polarized Slow Neutron Beams ... 13
    1.3.1 Sources of Polarized Cold Neutrons 14
    1.3.2 Neutron Guides 14
    1.3.3 Polarizing Neutron Guides 16
    1.3.4 Multilayer Polarizing Monochromators 20
    1.3.5 Polarizing Supermirrors 21
    1.3.6 Systems Generating Polarized Neutron Beams 21

2. Precession of Magnetic Moments of Polarized Neutrons in a Magnetic Field 23
  2.1 Fundamentals of the Magnetic Resonance Depolarization Technique 24
  2.2 Measurements of the Neutron Magnetic Moment 25
    2.2.1 Survey of Experiments 25
    2.2.2 The Magnetic Resonance Neutron Spectrometer at Grenoble 26
    2.2.3 The Value of the Neutron Magnetic Moment 27
    2.2.4 The Sign of the Neutron Magnetic Moment 27
    2.2.5 Comparison with the Theory 27
  2.3 Search for the Electric Dipole Moment of the Neutron 28
    2.3.1 Fundamentals of the Method 28
    2.3.2 The Experiment on Measuring the Electric Dipole Moment of the Neutron 29

	2.3.3	The Result of Searching for the Electric Dipole Moment of the Neutron	34
	2.3.4	Other Experimental Methods in the Search for the Electric Dipole Moment of the Neutron	35
	2.3.5	Comparison with the Theory	36

3. Interaction of Polarized Neutrons with Polarized Nuclei 37
  3.1 Fundamentals of the Method 37
  3.2 Survey of Experiments 39
  3.3 Nuclear Precession of Neutrons 42

4. Anisotropy of Gamma Rays Emitted by Polarized Nuclei After Polarized Neutron Capture 44
  4.1 Fundamentals of the Experimental Method 44
  4.2 Survey of Experiments 47
  4.3 Coherent Interference of Spin States 49

5. Circular Polarization of Gamma Rays Emitted by Nuclei After Polarized Neutrons Capture 50
  5.1 Fundamentals of the Experimental Method 50
  5.2 Survey of Experiments 52

6. Quantum-Mechanical Symmetry Properties and Polarized Neutrons 55
  6.1 Quantum-Mechanical Conservation Laws 55
  6.2 Violation of P, C, and CP Parities in the Weak Interaction 57
  6.3 The Role of Polarized Neutrons in the Investigation of Quantum-Mechanical Symmetry Properties 58
  6.4 Some Aspects of the Weak Interaction Theory 58
  6.5 The Structure of the Nucleon-Nucleon Weak Interaction 60
  6.6 Enhancement Mechanisms of the Nucleon-Nucleon Weak Interaction 61

7. Decay of Polarized Neutrons 64
  7.1 Fundamentals of the Experimental Method 64
  7.2 Spatial Parity Violation 65
    7.2.1 Experiment 65
    7.2.2 Results 67
  7.3 Investigation of Invariance Under Time Reversal 68
    7.3.1 Experiment 68
    7.3.2 Results 70

8. Anisotropy of Gamma Rays Emitted by Nuclei After Polarized Neutron Capture ... 71
8.1 Fundamentals of the Experimental Method ... 72
8.1.1 Estimate of the Expected Enhancement of P-Odd Effects ... 72
8.1.2 Angular Distribution of Gamma Quanta Emitted by Nuclei After the Capture of Polarized Thermal Neutrons ... 72
8.2 Experimental Investigation of Spatial Parity Violation in Nuclear Interactions ... 74
8.2.1 Choice of Nuclei ... 74
8.2.2 Specifics of the Reactions ^{113}Cd$(\vec{n}, \gamma_0)$ and ^{117}Sn$(\vec{n}, \gamma_0)$ ... 74
8.2.3 First Attempt at Detecting a P-Odd Effect in (n, γ) Reactions ... 75
8.2.4 Experimental Discovery of the Effect ... 76
8.2.5 Other Experiments on the Analysis of P-Odd Effects in the ^{113}Cd$(n, \gamma)^{114}$Cd Reaction ... 78
8.2.6 A Study of the P-Odd Effect in the Reaction ^{117}Sn$(\vec{n}, \gamma_0)^{118}$Sn ... 79
8.2.7 An Attempt to Study P-Odd Effects in Light Nuclei ... 80
8.2.8 Analysis of P-Odd Effects in the Integral Spectrum of Gamma Quanta ... 81
8.2.9 Results of Studying the Anisotropy of Gamma Radiation Emitted by Nuclei After Polarized Neutron Capture ... 81
8.3 Experimental Investigation of Time Reversal Invariance in Nuclear Interactions ... 83
8.3.1 Introductory Remarks ... 83
8.3.2 Fundamentals of the Method ... 83
8.3.3 Experiments ... 85
8.3.4 Measurement Results for Angular Correlations of Gamma Quanta Emitted by Nuclei After Polarized Neutron Capture ... 87

9. Anisotropy of Beta Particles Emitted by Nuclei After Polarized Neutron Capture ... 89
9.1 Fundamentals of the Experimental Method ... 89
9.1.1 Polarization of Nuclei ... 89
9.1.2 Angular Anisotropy of the Beta Emission ... 91
9.1.3 Experimentally Measured Asymmetry of the Beta Emission ... 91
9.1.4 Nuclear Magnetic Resonance of Polarized Beta-Radioactive Nuclei and the Measurement of Nuclear Magnetic Moments ... 93

      9.1.5 Measurements of Quadrupole Moments of Nuclei 93
  9.2 Survey of Experiments 94
  9.3 Summary of Experimental Data 97

10. Anisotropy of Alpha Particles and Other Light Nuclei Emitted After Polarized Neutron Capture 98
  10.1 Fundamentals of the Experimental Method 99
  10.2 Survey of Experiments 100
      10.2.1 Selection of Nuclei 100
      10.2.2 Requirements for the Detector Unit of the Apparatus ... 101
      10.2.3 Experimental Systems 101
  10.3 Summary of Results 105

11. Anisotropy of the Angular Distribution of Fragments After Fission of Heavy Nuclei by Polarized Neutrons 106
  11.1 Survey of Experiments 107
      11.1.1 Specific Features of Fission Reactions 107
      11.1.2 Discovery of P-Odd Asymmetry in the Emission of Fission Fragments 107
      11.1.3 An Analysis of the Asymmetry Coefficient as a Function of Fragment Mass and Neutron Energy 109
      11.1.4 Asymmetry in Fission Neutron Emission 111
  11.2 Summary of Experimental Data 112
  11.3 Investigation of P-Even Angular Correlations 112
  11.4 Attempt at a Theoretical Interpretation 114

12. Spatial Parity Violation Effects in Neutron Optics 117
  12.1 Fundamentals of the Experimental Method 118
  12.2 Survey of Experiments 121
      12.2.1 Neutron Spin Rotation 121
      12.2.2 Transmission of Longitudinally Polarized Neutrons 123
  12.3 Summary of Experimental Data 126

References 129

Subject Index 137

1. Brief Summary of Experimental Techniques Required for Working with Polarized Slow Neutrons

This book is primarily devoted to describing and analyzing the experiments in which polarized beams of slow neutrons are used for solving fundamental problems of nuclear physics. Nevertheless for the sake of completeness, at least a brief discussion of the methods for working with polarized neutrons is necessary here.

The material summarized in this chapter on devices developed in recent decades will be of special interest to the reader. These devices include new types of spin flippers for achieving the mutual rotation of the vectors of the leading magnetic field strength and the neutron spin without introducing any substance into the neutron beam; neutron guides, including polarizing neutron guides, for transporting neutrons over large distances; multilayer polarizing monochromators; and polarizing supermirrors.

The table at the end of the chapter lists the characteristics of a number of systems for generating polarized neutron beams. The list is definitely incomplete, mostly describing the installations used to carry out basic research described in the chapters to follow.

1.1 Generation and Analysis of Polarized Slow Neutron Beams

1.1.1 Behavior of Neutron Spins in Magnetic Fields

As magnetic fields are often used to generate polarized beams of slow neutrons, we shall consider the behavior of neutron spins passing through a region with a magnetic field H. The equation of motion for the mean-value spin s in the nonrelativistic case [1] is

$$d\boldsymbol{s}/dt = (2\mu_\mathrm{n}/\hbar)\boldsymbol{s}\times\boldsymbol{H} , \qquad (1.1)$$

where μ_n is the magnetic moment of the neutron. This equation describes the Larmor precession of the spin s around the direction of H at a frequency $\omega_\mathrm{L} = \gamma_\mathrm{n} H$, where $\gamma_\mathrm{n} = 2\mu_\mathrm{n}/\hbar$ is the gyromagnetic ratio for the neutron.

Let us consider the behavior of the mean-value spin of a neutron passing through a magnetic field that rotates in the reference frame moving with the neutron. Let the orientation of the axes x, y, z of this reference frame remain constant and let the axis z be directed along the velocity vector of the neutron. We fix the spin orientation at the instant of time $t = 0$ by the following initial conditions:

$$s_x(0) = s_0, \quad s_y(0) = s_z(0) = 0 , \tag{1.2}$$

where s_x, s_y, s_z are the components of spin. The magnetic field rotating in the system of coordinates x, y, z can be written in the form

$$H_x = H_0 \cos(\omega_0 t), \quad H_y = H_0 \sin(\omega_0 t), \quad H_z = 0 . \tag{1.3}$$

The solution to (1.1), satisfying the conditions (1.2) and (1.3), can be found in [2]. We shall give the values of the components of the spin vector s in the particular case of the neutron after it crosses a magnetic field rotated by 180°. If L is the length of the region within which H rotates through 180°, and v is the neutron velocity, the angular velocity of rotation of the vector H (in the reference frame fixed to the neutron) is $\omega_0 = \pi v/L$. Let us introduce the frequency of the Larmor precession of the neutron spin in a magnetic field of strength H_0: $\omega_L = \gamma_n H_0$. As shown in [2], the angle of rotation of the spin is determined by the ratio $k = \omega_L/\omega_0 = \omega_L L/\pi v$.

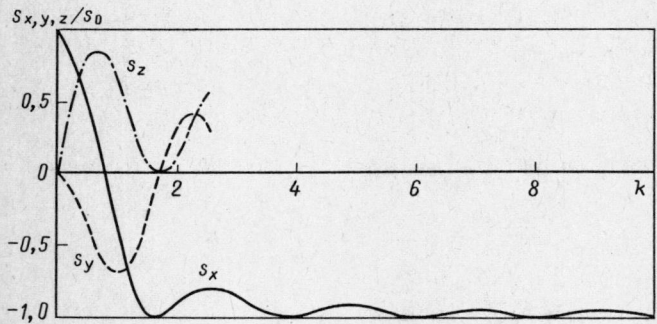

Fig. 1.1. The components of the average neutron spin s as functions of $k = \omega_L/\omega_0$, after the passage of the neutrons through a magnetic field rotating spins by 180°

Figure 1.1 plots s_x, s_y, and s_z as functions of k. Let us analyze the behavior of s_x. As can be seen in Fig. 1.1, at large k, the component s_x reverses the sign without appreciably changing the magnitude; i.e., the spin of the neutron rotates adiabatically (see Sect. 1.2.2), together with the vector H, through 180°. But if k is sufficiently small, the neutron passing through a magnetic field rotating through 180° will practically preserve the magnitude and orientation of its spin. The direction of the spin will then be opposite to that of the magnetic field; i.e., the so-called nonadiabatic spin flipping (spin reversal) will take place [3].

The condition of adiabaticity of spin rotation together with the magnetic field is, therefore, written as

$$k \gg 1 \quad \text{or} \quad \omega_L \gg \omega_0 ; \tag{1.4}$$

i.e., the angular speed of rotation of H must be small compared with the precession frequency of the neutron spin in this field. The condition of nonadiabaticity of the neutron passage through a region with a magnetic field is the reversed inequality:

$$k \ll 1 \quad \text{or} \quad \omega_L \ll \omega_0 \,. \tag{1.5}$$

1.1.2 Schematic Diagram of a System for Generating and Analyzing Polarized Neutrons

Before we begin describing the system for generating a polarized neutron beam, we must give a quantitative definition of beam polarization. A beam of spin-1/2 particles (in units of $\hbar$) is said to be polarized if the quantity

$$P_n = (N_+ - N_-)/(N_+ + N_-) \tag{1.6}$$

is nonzero; here N_+ is the number of particles in the beam with spin projections parallel to some physically favored direction in space (e.g., the direction of the magnetic field), and N_- is the number of particles in the same beam whose spin projections are antiparallel to this direction. Only two spin orientations are possible because the neutron spin is $s = 1/2$. The quantity P_n is called the beam polarization. Usually, it is assumed that $N_+ \geq N_-$, so that $0 \leq P_n \leq 1$. A beam with $P_n = 1$ is said to be totally polarized, while a beam with $P_n = 0$ is described as unpolarized.

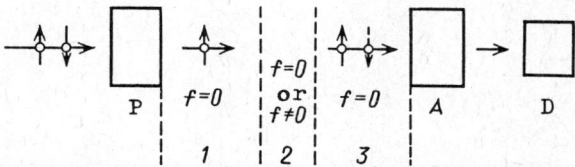

Fig. 1.2. Schematic diagram of the system for obtaining and analyzing the polarized neutron beam

A typical schematic diagram of a system for generating and analyzing a polarized neutron beam is given in Fig. 1.2. An unpolarized neutron beam is incident on a polarizer P. The degree of polarization of the beam emerging from the polarizer will be denoted by P_{n1}. The beam then passes through an analyzer A and is detected by a neutron counter D. The polarizing properties of the analyzer with respect to the incident unpolarized beam will be characterized by a quantity P_{n2} called the polarization efficiency; in certain cases, it may equal the degree of polarization of the neutron beam emerging from the polarizer, but in general, it is different [4].

In order to preserve the orientation of neutron spins in the space between the polarizer and the analyzer (or the specimen), a leading magnetic field is created. This field suppresses the depolarization due to the possible Larmor precession of neutron spins in perturbing stray magnetic fields. The leading field must be either parallel or antiparallel to the direction of polarization. As a rule, an apparatus for the generation and analysis of polarized neutron beams is assembled so as to make regions 1 and 3 (see Fig. 1.2) "adiabatic" (from the standpoint of the rotation of neutron spins), while region 2 may be either "adiabatic" or "nonadiabatic".

In order to characterize the degree of nonadiabaticity (or adiabaticity) of a region with the leading magnetic field, we introduce the probability f for the neutron to change the orientation of its spin to the opposing orientation after passing through this region. We refer to f as the spin flip probability. Then, $f = 0$ in adiabatic, and $f \neq 0$ in nonadiabatic regions; if the nonadiabatic region completely depolarizes the beam, then $f = 1/2$, and if it completely reverses the spins with respect to the leading field, then $f = 1$.

1.1.3 Measurement of Beam Polarization

Typically, one of the following two methods is used to measure the degree of beam polarization produced by a polarizer: double reflection or a shim method (a shim is a thin plate of nonmagnetized iron). The two methods are presented in detail in [4]. Here, we give only the fundamentals of the methods and the final formulas. In the double reflection method, a nonadiabatic region is formed between the polarizer and the analyzer, with a spin flip probability of $f = 1$ (or nearly 1); i.e., a region is formed that completely reverses the spins with respect to the leading field. In order to find polarization, we need to measure the so-called polarization ratio

$$R' = n_2/n_2' , \qquad (1.7)$$

where n_2 is the detector counting rate in the absence of a nonadiabatic region between the polarizer and analyzer, and n_2' is the counting rate in the presence of a nonadiabatic region. If the sensitivity of the detector varies as $1/v$, where v is the velocity of neutrons, then the counting rate is proportional to the density of neutrons in the beam. If the sensitivity of the detector is independent of v (the so-called "black" detector), its counting rate is proportional to the neutron flux density.

Now we can write

$$P_{n1}P_{n2} = (R' - 1)/(R' + 1) . \qquad (1.8)$$

With the shim method, a nonadiabatic region with $f = 1/2$ is created, i.e., a region that completely depolarizes the transmitted beam. To achieve this, the beam is passed through a shim. Multiple nonadiabatic spin reversals occurring on passing through the domains of the ferromagnetic plate with randomly oriented magnetization result in the depolarization of the beam. The so-called shim ratio R is measured to calculate the beam polarization

$$R = n_2/n_{2s} , \qquad (1.9)$$

where n_2 is the counting rate of the neutron detector without a shim and n_{2s} is the counting rate with the shim introduced. Therefore,

$$P_{n1}/P_{n2} = R - 1 . \qquad (1.10)$$

Note that using the shim requires that we make corrections for the effects of nuclear scattering and absorption and for the magnetic small-angle scattering due to the refraction of neutron trajectories on the passage across domain boundaries.

The absorption of neutrons in the shim due to nuclear interactions is easily taken into account in measuring the counting rate n_2 by placing the shim between the analyzer and the detector (as close to the analyzer as possible).

Several methods described in [4] have been suggested for taking into accout the small-angle scattering; the simplest of them is the two-shim technique. A method of measuring the depolarizing properties of shims can also be found in [4].

The quantity yielded by any of these methods is the product $P_{n1}P_{n2}$. In order to find each of the factors, we need to know the relation between P_{n1} and P_{n2}. If the analyzer and polarizer are identical (i.e., both their physical properties and the conditions in which they are placed are identical), it can be shown that $P_{n2} \geq P_{n2}$ [4]. In the general case of a nonidentical polarizer and analyzer, $P_{n2} \geq cP_{n1}$. The factor c can be found by comparing the polarizing efficiencies of the polarizer and analyzer [4]. They are placed successively as analyzers in the same polarized beam generated by a third polarizing device, and the products $P_{n3}P_{n1}$ and $P_{n3}P_{n2}$ are measured. If the polarizer and analyzer are identical, these products are equal, and $c = 1$. With $P_{n1}P_{n2}$ and c known, we can determine the bounds on P_{n1} and P_{n2} [4]. The methods of finding the probability f of spin reversal by a nonadiabatic region are discussed in detail in [4].

1.2 Experimental Methods for Studying Polarized Slow Neutron Beams

1.2.1 Collimators, Filters, Detectors

Polarized beams of slow neutrons are produced at present at research nuclear reactors. It will thus be useful to discuss the experimental techniques used in reactor facilities.

The neutron beam incident on the polarizer must be collimated. The collimation is characterized by the angle between the longitudinal axis of the collimator and the outermost trajectory in the neutron beam. The efficient collimation of the beam incident on the polarizer is determined by the dimension of the opening at the input of the collimator inside the reactor shield, and by polarizer dimension.

Beam collimation of several degrees of arc is needed for the creation of polarized beams by means of diffraction on crystals, and the collimation of several arc minutes is needed to create beams by reflection from mirrors.

Beam divergence is reduced by diaphragms which inevitably lower the beam intensity. Thermal neutrons are absorbed by diaphragms made of a thin (about 1 mm thick) cadmium layer. The absorption of the resonance and fast neutrons and of γ-quanta occurs in collimating blocks composed of steel, lead, boron carbide, boric acid, and cadmium. Recently the advent of neutron guides made it possible to extract from reactors polarized neutron beams with no admixture of fast neutrons and gamma quanta (see Sect. 1.3).

The neutron flux density in a reactor beam is evaluated in terms of the geometric dimensions of the collimator which forms the reactor beam. If the collimator opening of area S cm^2 is located at a point where the neutron flux density is Φ_0 neutrons/(cm$^2\cdot$s) (the "luminous" neutron surface), then the neutron flux density Φ at a point of interest located at a distance of l cm from the luminous neutron surface is

$$\Phi = \Phi_0 S / 4\pi l^2 \ . \tag{1.11}$$

The horizontal angular divergence of the beam is reduced by inserting thin vertical plates into the collimator (the so-called Soller type collimator).

In collimators with small angular divergence, it is often necessary to take into account the total reflection of neutrons from the walls of the collimator slits, since it results in enhanced beam divergence. In order to avoid this effect, either slit walls are made of materials with nearly zero scattering length for slow neutrons, or thin strips made of neutron-absorbing material are inserted into the slits [4].

Cold polarized neutrons are obtained by placing neutron crystal filters in front of the main collimator. Neutrons whose wavelength is greater than the cutoff value $\lambda_c = 2d$, where d is the maximum interplanar spacing of the crystal lattice, are transmitted through the filter at practically unattenuated intensity. Neutrons with shorter wavelengths are removed from the beam by the Bragg reflection and absorbed in the cadmium shield of the filter.

The material of the filter must have small cross sections of absorption and noncoherent scattering for the transmitted neutrons, and a sufficiently high scattering cross section for the scattered neutrons. The most suitable materials for filters are graphite, beryllium, and beryllium oxide (cutoff wavelengths being 0.669 and 0.44 nm, respectively). The attenuation of the cold neutron flux in the filters described above is chiefly caused by thermal inelastic scattering. The cross section of this scattering being strongly temperature-dependent, further enhancement of transmission can be achieved by cooling the filter. (For new techniques of producing polarized neutron beams, see Sect. 1.3.)

Polarized neutrons are detected either by proportional counters filled with gaseous BF_3 (often with boron enriched in the ^{10}B isotope) or ^3He, or by scintillating glasses containing ^6Li. Thus, a counter developed at the Laue-Langevin Institute at Grenoble in France [5] consists of two glasses containing lithium (^6Li), in close contact with two photomulitpliers; the counter successfully detected polarized neutron fluxes up to $5 \cdot 10^6$ neutrons/s. Still higher neutron fluxes are detected by an integral technique first suggested in [6].

Neutron counters are covered by an appropriate shield and the whole assembly is moved in the horizontal plane perpendicularly to the beam, to the accuracy of 0.5 mm. A cadmium or boron-cadmium collimator with variable slit width is mounted at the entrance to the counter.

1.2.2 Adiabatic Spin Rotation

The condition (1.4) of the adiabaticity of neutron spin rotation in a magnetic field can be rephrased as follows: the distance covered by a neutron during one period of its Larmor precession must be small in comparison with the distance over which the magnetic field strength vector is reversed. The distance covered by thermal neutrons during one precession period is $64/H_0$, where H_0 is the magnetic field strength in A/cm. As follows from Fig. 1.1, neutron spins are adiabatically rotated together with the direction of the leading magnetic field if $k \geq 6$.

The leading field along the path of the polarized neutron beam is produced either by permanent magnets or by electromagnets, or by passing a direct current through conductors placed along the beam. Typically, the leading magnetic field strength is in the range 5–240 A/cm.

In [7], the direction of polarization of the beam of thermal neutrons was flipped at a frequency of 10 Hz by using a low-inertia rotating magnet made of transformer steel sheets. It was shown [8] that the total spin reversal by this magnet occurs over not more than 20 ms from the moment of current reversal in the magnet coils.

While the spins of thermal neutrons can be rotated over a relatively short distance even by stray fields of magnets rotated by 90°, this method is insufficient for resonance neutrons. The adiabatic spin rotation of resonance neutrons was achieved in [9,10] by passing the beam through magnets rotated by a relatively small angle with respect to each other. Neutron spins were reversed by mechanically reversing the helicity of this magnetic guide.

It should be noted in conclusion that a smooth rotation of the polarization vector of a polarized neutron beam is achieved rather easily, with no loss in the absolute value of polarization.

1.2.3 Spin Flippers

Experiments with polarized neutrons often require that the mutual orientation of the spin s and the leading magnetic field H_0 be changed. Experimental devices that implement the mutual reversal of s and H_0 are called spin flippers. They can be classified into three types: (1) resonance spin flippers with oscillating field, based on the spin resonance technique; (2) spin flippers using the method of rapid reversal of the leading magnetic field; and (3) spin flippers using the method of adiabatic passage of neutrons through a resonance.

Let us discuss the techniques on which the functioning of these devices is based.

Spin Resonance Technique

This method, based on the resonant absorption by neutrons of the energy of an oscillating magnetic field, was already employed in the earliest experiments on measuring the magnetic moment of the neutron [11–13]. A weak rf oscillating field $H_1 \cos(\omega t)$ is applied at right angles to the vector s and to the direction of the uniform leading magnetic field.

The probability f of spin reversal with respect to the direction of the leading magnetic field H_0 during a time t of passing through the rf field is given by the formula [3, 11, 13]

$$f = \frac{(\omega/\omega_L)^2 \sin^2\{(\mu_n H_1 t/2\hbar)[\omega/\omega_L + (2\Delta H/H_1)^2]^{1/2}\}}{\omega/\omega_L + (2\Delta H/H_1)^2} . \qquad (1.12)$$

In this expression, $\Delta H = H_0 - H_0^*$, where $H_0^* = \omega/\gamma_n$.

As follows from (1.12), two conditions must be satisfied for the spins to be flipped:

$$\Delta H = 0 \; ; \quad \text{i.e.,} \quad \omega = \omega_L = \gamma_n H_0 \; , \qquad (1.13)$$

$$\mu_n H_1 t/2\hbar = n\pi/2 \; , \qquad (1.14)$$

where $n = 1, 3, 5, \ldots$. If both conditions are met, the probability of spin reversal is $f = 1$.

The time of flight through the rf field is $t = L/v$, where L is the length of the coil producing the field and v is the neutron velocity, so that the spin flip probability depends on the velocity of the neutron. If the conditions (1.13) and (1.14) are satisfied for a neutron with wavelength λ_0, i.e., $f(\lambda_0) = 1$, then the probability of spin flip for a neutron with a different wavelength λ (for $n = 1$) is

$$f(\lambda) = \sin^2\{(\pi/2)(\lambda/\lambda_0)\} . \qquad (1.15)$$

The probability of spin reversal in the spin resonance method is very sensitive to the energy of the neutrons. This factor restricts the application of resonance spin flippers with oscillating magnetic field: They are used only with monoenergetic polarized neutron beams. The direction of polarization is reversed by switching on the rf coil.

We note in conclusion that the oscillating-field resonance spin flippers readily produce a spin flip probability $f = 0.98\text{--}0.99$. Among the shortcomings of this method are the already mentioned strong dependence on neutron energy and the need to highly stabilize the frequency and amplitude of the oscillating magnetic field and to make the leading magnetic field sufficiently uniform in the entire region in which neutrons pass through a radio-frequency coil.

Method of Rapid Reversal of the Leading Magnetic Field

In contrast to the foregoing technique, this method is suitable for nonmonoenergetic neutron beams; hence, its wide popularity.

In this technique, the region in which the orientation of the leading magnetic field is reversed is narrowed down so as to make the time of flight of neutrons through it less than the Larmor precession period. The nonadiabaticity condition (1.5) being satisfied, a neutron passing through the narrow region of field reversal retains the orientation of its spin while that of the leading field changes to the opposite direction.

The nonadiabatic flipping of neutron spins is analyzed in [3, 14]. According to [14], a neutron passing with a velocity v along the axis z through a region of magnetic field reversal of length L, in which the field changes so that

$$H_z = \text{const}, \quad H_x = 0, \quad \partial H_y/\partial t = \text{const}$$

(beyond the L-long region the neutron spin points along the axis y, and t is the time in the reference frame moving with the neutron), has the following probability of spin flip after passing through this nonadiabatic region:

$$f = \left[\exp -\frac{\pi \mu_n H_z^2}{\hbar v \partial H_y/\partial z}\right] = \exp\left[-\frac{\pi \omega_L}{2\omega_0}\right], \quad \text{where} \tag{1.16}$$

$$\omega_L = \gamma_n H_z, \quad \omega_0 = (\partial H_y/\partial t)/H_z$$

is the angular velocity of rotation of the field strength vector. As follows from (1.16), the spin flip probability is the closer to unity, the steeper the field gradient in the nonadiabatic region and the smaller the component of the field perpendicular to the direction of the neutron spin.

The fundamentals of this technique are presented in detail in [4], so that only its modifications will be discussed here. Beginning with the experiment by *Dabbs* et al. [15], the sharp reversal of the leading magnetic field is achieved by using a thin metal foil carrying dc current. A design of the foil spin flipper and the values of spin reversal probability obtained with it are given in a paper by *Gulko* et al. [16] (see also [4]).

It was found that the efficiency of polarization reversal strongly depends on the location of neutron passage through the foil but is independent of its thickness. Neutrons passing along the symmetry axis of the foil undergo total polarization reversal. As the beam moves away from the symmetry axis, the polarization reversal becomes incomplete. It was shown in [17] that this effect is caused by the magnetic field components perpendicular to the foil surface. This paper also gave the formulas for the efficiency of polarization reversal, compared the results of calculations with the experiments, and discussed the principles and implementation of a method of compensating the perpendicular component of the magnetic field; the method proved to substantially enlarge the region of high efficiency of polarization reversal.

Abrahams et al. [18] replaced the foil with a device made of copper wire (see [4]). A similar wire spin flipper was later used by *Abov* et al. [7, 19]. The probability of spin flipping for a polarized nonmonoenergetic neutron beam reflected by cobalt mirrors was found to be $f = 0.98 \pm 0.04$. Nonadiabatic (i.e., using the nonadiabticity condition) wire spin flippers have an advantage over foil spin flippers in that they require lower currents (about 0.5 A instead of several hundred amperes) and can be switched on and off more easily.

A shortcoming of foil and wire spin flippers using the method of rapid reversal of the leading magnetic field is the introduction into the polarized neutron beam of some substance (foil or wire) which gives rise to additional gamma quanta due to the radiative capture of neutrons by the substance in the beam and to the small-angle scattering of neutrons by this substance.

In order to eliminate this shortcoming, spin flippers were proposed which form the region of rapid reversal of the leading field direction without any substance introduced into the beam. Such a spin flipper was first used by *Drabkin* et al. [20]. It consisted of two coaxial coils, with oppositely directed currents, 60 mm in diameter, placed 400 mm apart. A beam of polarized thermal neutrons was sent along the z axis of the coils. With the outer magnetic fields suppressed by a magnetic screen, the magnetic field on the z-axis of the coils had, owing to axial symmetry, only the z component and reversed its sign at the middle point between the coils, provided the coil currents were equal. The spin flipper of this design provided the spin flip probability $f = 1$ for a neutron beam with a cross section of 18×3 mm. However, the flip probability decreased for beams with a larger cross section.

An attempt to extend the region of efficient spin reversal in a spin flipper with axial geometry was made in [21], where the authors considerably increased the size of the device. The coil diameter was 30 cm and the distance between the coils was 85 cm. The probability of spin reversal for polarized cold neutrons (at an average velocity of 1100 m/s) equalled 1 at the beam axis, but already dropped to 0.95, 2.5 cm from the axis.

Korneev [22] suggested a spin flipper producing a more extended region of total polarization reversal for beams with a cross section elongated in one direction (total reflection from magnetized mirrors and polarized neutron guides form beams of precisely this shape). The design of this spin flipper and the behavior of the *s* and *H* vectors are shown in Fig. 1.3. Two rectangular coils with oppositely directed currents were arranged in the xy plane perpendicular to the beam axis. Two compensating coils were used to suppress the H_x component of the stray magnetic fields of the polarizer and analyzer magnets. The

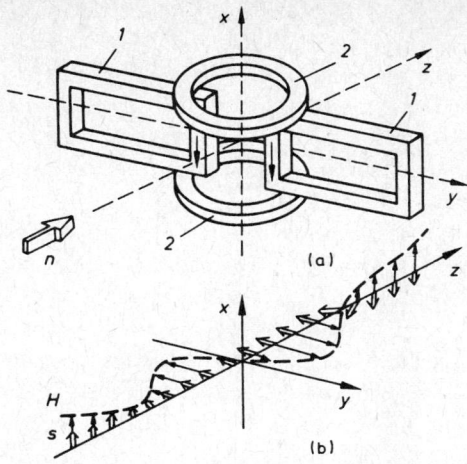

Fig. 1.3. General view of coil arrangement in Korneev's spin flipper [22] (a), and (b) the behavior of spin vectors (*double arrows*) and magnetic field **H** (*single arrows*) with the spin flipper switched on: *(1)* current frames, *(2)* compensating coils; the neutron beam is elongated along the x-axis, and the neutrons with spins pointing along the y-axis have velocity along the z-axis; the arrows show the direction of currents in the mode of nonadiabatic reversal of neutron spins with respect to **H**

neutron beam elongated along the x-axis, with spins aligned along the y-axis, entered the region of rapid reversal of **H** (the xy plane) where the nonadiabatic flipping of neutron spins with respect to the magnetic field vector took place (see Fig. 1.3b).

The spin flipper was tested on a beam of polarized thermal neutrons, the width and height of the beam cross section being $\Delta y = 0.5$ mm and $\Delta x = 22$ mm, respectively. The maximum probability of spin reversal was $f = 0.9993 \pm 0.0001$. The experiment confirmed the quadratic dependence of spin flip probability on the y coordinate of the beam centroid, and the linear dependence on wavelength λ,

$$f(\lambda, y) = 1 - c\lambda y^2 \partial H_y / \partial z , \qquad (1.17)$$

where $\partial H_y / \partial z$ is the value of the gradient of the leading magnetic field. The validity of formula (1.16) for the spin flip probability when passing through a nonadiabatic region with zero magnetic field, from which formula (1.17) was derived, was thereby confirmed.

An obvious advantage of this flipper design is its short extension along the beam axis, and its shortcoming is the restriction of the beam width.

Mezei [23] proposed a spin flipper for monoenergetic neutrons, based on the method of leading field reversal. This spin flipper consists of two rectangular coils, each of thickness d, placed one after the other along the beam (Fig. 1.4). The coils produce oppositely oriented horizontal magnetic fields of strength H, equal to the strength of the vertical leading magnetic field H_0. If the coil

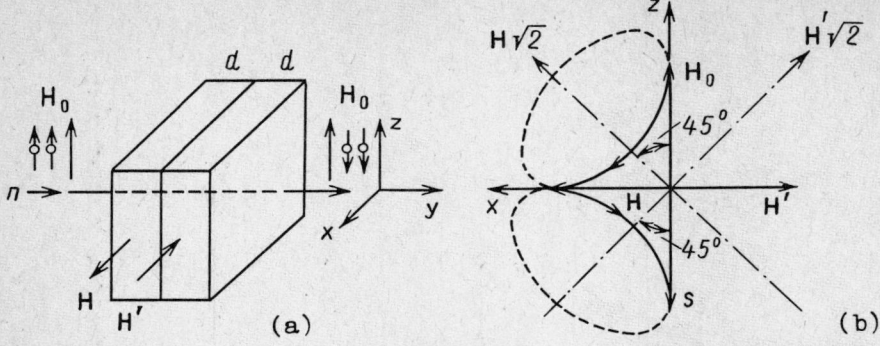

Fig. 1.4a,b. The principle of functioning of the Mezei spin flipper [23]: (a) arrangement of coils, (b) action of magnetic fields; n shows the polarized neutron beam; $H_0 = H = H'$

thickness satisfies the condition

$$d/v = \pi/\omega_L = \pi/\gamma_n H_0 \sqrt{2} \,, \tag{1.18}$$

where ω_L is the Larmor frequency of spin precession in the total field $H_0\sqrt{2}$ within the coils, then the neutron spins rotate in each coil by 90° and pass nonadiabatically across the field reversal boundary between the coils; as a result, they emerge from the second coil being oriented oppositely to the leading field H_0.

Experiments [24] confirmed the effectiveness of the Mezei spin flipper. The probability of spin flipping was measured in a spin flipper with parameters $d = 1.5\,\text{cm}$, $\lambda = 0.14\,\text{nm}$, $H_0 = 18\,\text{A/cm}$, the coil dimensions being $8\times 4\,\text{cm}$. The device yielded f up to 0.997.

The Mezei spin flipper served as a basis for developing new interesting devices: a spin echo spectrometer [23], and a spin flip chopper of polarized neutron beam [24]. Both instruments can be used for studying the processes of inelastic scattering of slow neutrons.

Van Laar et al. [25] reported the design and testing of a single-coil dc spin flipper. The effect of the reversed field of the coil in its vicinity was taken into account, and the characteristics of spin flipping were numerically calculated. The experiments on a beam of monoenergetic neutrons with $\lambda = 0.1113\,\text{nm}$ at the high-flux reactor in Petten (the Netherlands) indicated that the spin flip probability f reached 0.998.

Method of Adiabatic Passage of Neutrons through Magnetic Resonance

A spin flipper based on this method was first applied in neutron physics in [26]. It originates from the work on nuclear magnetism where such a device is used to flip the nuclear magetization vector [27, Chap. II].

Briefly, the method of adiabatic passage as applied to a polarized beam of thermal neutrons is as follows [28]. A nonuniform (increasing or decreasing) stationary magnetic field H_0 and a magnetic field H_1 perpendicular to

H_0, rotating at a frequency ω, are created along the path of a neutron beam. The expression for the effective magnetic field strength in the reference frame comoving with the neutron and rotating at the same frequency ω around H_0 is

$$H_e = [H_0(t) - \omega/\gamma_n]k + H_1 i \; , \tag{1.19}$$

where k and i are the unit vectors in the directions of H_0 and H_1, respectively, and $H_0(t)$ is the field strength H_0 in the moving reference frame as a function of time. The effective field constantly changes its direction in space. The system is so adjusted that the resonance condition $\omega = \omega_L = \gamma_n H_0$ is met in some region within the spin flipper. The effective field in this region is minimal and equal to H_1. If H_0 changes sufficiently slowly, the neutron spin projection on the direction of the effective field is conserved ("adiabaticity theorem" [27]). If the initial value of H_0 is much greater than the resonance field, so that the effective field is practically parallel to H_0, and the final value of H_0 is much less than the resonance field, then the neutron spin which was first parallel to H_0 will remain parallel to H_e and therefore, will finally be antiparallel to H_0. As shown in [27], the condition of polarization reversal is the inequality

$$dH_0/dt \ll \gamma_n H_1^2 \; . \tag{1.20}$$

The schematic of a system at the B.P. Konstantinov Institute of Nuclear Physics in Leningrad (LINP) using an adiabatic spin flipper of polarized thermal neutrons is given in Chap. 12 (see Fig. 12.2). The field H_0 was produced by dc coils, and H_1, by an rf coil.

The advantages of adiabatic spin flippers are obvious: (1) the efficiency of spin flipping is independent of neutron velocity if this velocity is less than the maximum value for which (1.20) is satisfied; (2) the uniformity of the leading field H_0 is not required; (3) the collimation of the neutron beam is not as essential as in the other methods; and (4) no substances are introduced into the neutron beam within the spin flipper.

1.3 New Methods of Generating Polarized Slow Neutron Beams

Standard, commonly employed methods of generating polarized neutron beams are presented in detail in [4]. These methods are: neutron diffraction, total reflection from magnetized ferromagnetic mirrors, and transmission through magnetized and polarized targets. New reviews and monographs devoted to these methods have been published in [29], where the reader will find a detailed bibliography. Here we shall dwell only on the methods developed in the last decade for generating, transporting, and analyzing polarized neutron beams.

1.3.1 Sources of Polarized Cold Neutrons

In fundamental research with polarized neutrons, it is necessary to optimize not only the polarization $P(\lambda)$ but also the total neutron flux $J(\lambda)$ and the time of interaction of a polarized neutron with the target (or with the detecting part of the system). The time of interaction is proportional both to the length of the target (or of the system) and to the wavelength λ of the incident neutrons, so that it is preferable, other conditions being equal, to have neutrons with longer wavelengths. The accuracy with which an effect due to polarized neutrons can be measured is determined by the quantity

$$X = \int P^2(\lambda) J(\lambda) \lambda \, d\lambda \ . \tag{1.21}$$

In order to increase X, it is useful to obtain neutrons for the formation of polarized beams from a specially cooled zone in the reactor, called the cold neutron source. The methods of designing cold neutron sources are described in [30].

One of the most efficient cold sources functions at the high-flux ILL reactor [31]. A thin-walled aluminum vessel 38 cm in diameter, containing 25 liters of boiling deuterium at 25 K, is placed in a heavy-water moderator. The neutron spectrum in this source has its maximum at $\lambda = 0.5$–0.6 nm. The source yields a gain factor of 30 in intensity of neutrons at $\lambda = 0.6$ nm as compared with the spectrum of thermal neutrons. Neutron guides which transport cold neutrons over large distances from the reactor core are located in the immediate vicinity of the source.

1.3.2 Neutron Guides

These devices for transporting neutrons over large distances from the source are based on the total reflection of neutrons at the inner walls of neutron guides. A specailly created curvature of neutron guides makes it possible to eliminate background interference by fast neutrons and gamma quanta present in the neutron source.

The first neutron guides were installed at the FRM reactor in Munich [32]. Their principal characteristics are described in [33]. These guides utilize the total reflection of slow neutrons from polished surfaces.

The penetration of a neutron wave into a coherently scattering medium is accompanied by the interference of the incident and scattered waves, so that the wave vector of neutrons propagating across the interface between two media is changed. If noncoherent processes are neglected, the propagation of neutrons in the wall medium is described by introducing its refractive index. As in optics, the refractive index of a medium for a radiation incident from a vacuum is defined by the ratio

$$n = \cos\theta / \cos\theta' = k'/k \ , \tag{1.22}$$

where θ is the outer glancing angle at the plane interface when neutrons are incident on the wall from the vacuum; θ' is the inner glancing angle; k is the absolute value of the wave vector in the vacuum; and k' is the value of the wave vector in the medium.

The refractive index of a nonmagnetic medium is [34]

$$n = 1 - (\lambda^2 N/2\pi)b_N , \tag{1.23}$$

where λ is the neutron wavelength in the vacuum, N is the number of nuclei per $1\,\text{cm}^3$ of the scattering substance, and b_N is the coherent length of nuclear scattering on the bonded nuclei of the medium. For most nuclei, the coherent scattering lengths are positive, so that $n<1$, and $1-n\approx 10^{-6}-10^{-5}$. A medium with $b_N>0$ thus has a lower optical density for neutrons than a vacuum has, and hence, the effect of total reflection can arise for neutrons at the vacuum-medium (in fact, air-medium) interface. The maximum (critical) glancinig angle at which total reflection still occurs is found from the equality

$$\theta_c \approx \sin\theta_c = \sqrt{1-n^2} \approx \sqrt{2(1-n)} \approx 10^{-3}-10^{-2} . \tag{1.24}$$

From (1.23) and (1.24) we find that the critical angle is

$$\theta_c = (\lambda/\sqrt{\pi})\sqrt{Nb_N} . \tag{1.25}$$

This formula yields $\theta_c \approx 10'$ for $\lambda \approx 0.2$ nm. The reflection coefficient equals unity for glancing angles below the critical value, and steeply decreases for $\theta>\theta_c$.

The depth of penetration of neutrons into the scattering medium in the total reflection mode is very small. The neutron wave intensity reduces by a factor of e over a distance

$$d \approx \lambda/2\pi\,\theta_c \approx 10^{-6}\,\text{cm} . \tag{1.26}$$

The small depth of penetration in the total reflection of neutrons signifies very stringent requirements for the smoothness of the reflecting surface of neutron guides.

As shown in [33], a straight rectangular neutron guide of length L, slit height b, and slit width a, having well-polished inner walls, yields a gain factor G in the flux of transported neutrons as compared with a neutron guide with nonreflecting walls (for $\theta_c > a/L$ and $\theta_c > b/L$):

$$G = 4\theta_c^2 L^2/ab . \tag{1.27}$$

Thus, a copper neutron guide with $L = 7\,\text{m}$, $ab = 10\,\text{cm}^2$, has $G = 40$ for neutrons with $\lambda = 1$ nm ($\theta_c \approx 0.014$ rad).

As has been said, it is preferable to work with curved neutron guides (Fig. 1.5). The geometric parameters of a curved neutron guide are its curvature radius ϱ, length L, and channel width a. For all neutrons to undergo at least one reflection at the channel walls, L must be greater than the direct visibility

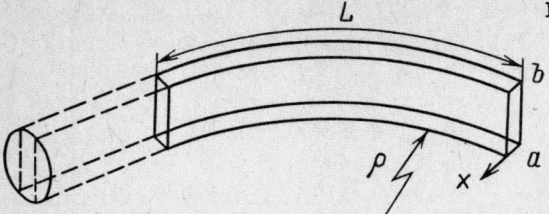

Fig. 1.5. The curved neutron guide

distance L_1 which, for $a \ll \varrho$, is given by the formula

$$L_1 = \sqrt{8a\varrho} \ . \tag{1.28}$$

This quantity is a function of another geometric parameter, the so-called characteristic angle γ of the neutron guide:

$$\gamma = \sqrt{2a/\varrho} = L_1/2\varrho \ . \tag{1.29}$$

The neutron guide axis deflects from the direction of the original beam to a distance $x = L^2/2\varrho$. In the case of a curved neutron guide, the gain in flux is somewhat diminished:

$$G = 8\theta_c^2 L^2/3ab \ . \tag{1.30}$$

The physical properties of the reflecting walls of a neutron guide depend on the characteristic neutron wavelength λ^* at which the critical angle θ_c becomes equal to the characteristic angle γ.

The first curved neutron guide at the FRM reactor was made of a circular copper pipe, 3.4 cm in diameter, 7 m long, with a curvature radius of 310 m. The gain factor G for neutrons with energy 10^{-3} eV was 35–50, as compared with a straight pipe with nonreflecting walls; the background level was significantly reduced [33].

At present, the neutron guides of the ILL reactor in Grenoble are up to 100 m long and have curvature radii 25 m–27 km.

Slow neutron guides attenuate the flux of slow neutrons by only 0.5–1 % per 1 m of guide length.

1.3.3 Polarizing Neutron Guides

If the walls of a neutron guide are made of polished magnetized ferromagnetic material, it becomes possible to polarize the transmitted beam. The first such devices were the multislit reflecting collimators designed by *Abrahams* and *Stecher-Rasmussen* with coworkers [35, 36] at the reactors in Risö (Denmark) and in Petten (the Netherlands).

In order to take into account the scattering medium of the magnetized ferromagnetic substance, the coherent length b_m of the magnetic scattering of neutrons by atoms is introduced into the formula for the refractive index [34]:

$$n_\pm = 1 - \lambda^2 N(b_N \pm b_m)/2\pi \ . \tag{1.31}$$

The plus and minus signs refer to the cases of the neutron spin being parallel and antiparallel, respectively, to the direction of magnetization of the mirror (the magnetic induction vector lies in the plane of the mirror).

The coherent magnetic scattering length b_m is an analog of the coherent nuclear scattering length b_N. However, nuclear scattering of thermal neutrons occurs on the nuclei of the scattering medium, while magnetic scattering occurs on the atoms of this medium. To be precise, neutrons are magnetically scattered because of the interaction between the magnetic moment of the neutron and that of the atom.

As shown in the theory of interaction between neutrons and the atoms of a ferromagnetic material (see [4]), the length b_m in the case of scattering on a plane mirror (the scattering vector is perpendicular to the mirror surface, and the vector H is parallel to it) can be given in terms of the magnetic induction B_s at total magnetic saturation of the ferromagnetic, neutron energy E, neutron wavelength λ, and the number N of atoms per unit volume of the scattering medium:

$$b_m = \pm |\mu_n| \frac{B_s - H}{E} \frac{\pi}{\lambda^2 N} \ . \tag{1.32}$$

The expression (1.32) includes the difference $B_s - H$ because of the continuity of the tangential components of H across the mirror-air interface.

The expression for the refractive index (1.31) then becomes

$$n_\pm = 1 - \frac{\lambda^2 N}{2\pi} b_N \mp |\mu| \frac{B_s - H}{2E} \ . \tag{1.33}$$

The validity of this formula was first checked experimentally in [37].

The critical angle is now given by the formula

$$\theta_c = \sqrt{\frac{\lambda^2 N}{\pi} b_N \pm |\mu_n| \frac{B_s - H}{E}} \ . \tag{1.34}$$

It was shown [4] that a neutron beam with high polarization and the spins aligned parallel to the direction of mirror magnetization can be obtained if

$$b_m \geq b_N \ . \tag{1.35}$$

In this case, $n_- \geq 1$ and $n_+ < 1$, so that the neutrons with spins parallel to the direction of the magnetic field (the plus sign) meet the condition of total reflection, while no such reflection occurs for neutrons with the antiparallel orientation of spins (the minus sign).

The neutron energy being $E = h^2/2m\lambda^2$, where m is the neutron mass, the expression (1.35) does not contain E, and hence, is valid for nonmonoenergetic neutron beams. However, characteristic wavelengths can be indicated for the reflection of neutrons in both spin states. The characteristic wavelength is the minimal value of λ at which the reflection coefficient equals unity. The shorter the wavelength, the greater the fraction of the neutron spectrum participating in reflection at a given glancing angle.

The materials for which the condition (1.35) holds are cobalt and some of its alloys; for example, vanadium permendur (50 % Co, 48 % Fe, 2 % V). Thus, in cobalt under saturation $b_N = 2.5 \times 10^{-13}$ cm, $b_m = 4.6 \times 10^{-13}$ cm.

The calculated dependence of reflection coefficients for neutrons in both spin states, R_+ and R_-, and polarizations P_n on neutron wavelength are listed in [4] for several values of the glancing angle θ, for an ideal plane cobalt mirror with magnetic induction $B = 1.5$ T.

A curved polarizing neutron guide was designed, installed, and tested by *Berndorfer* [38] at the FRM research reactor in Munich. The neutron guide consisted of 20 mirror sections, 25 cm long each. In each section, two mirrors were placed so as to form a slit 57 mm high and 5 mm wide. Each section was at a small angle to its neighbors, so that the total curvature radius of the whole system was 430 m. A vacuum of 1.33 hPa was maintained within the neutron guide. The mirrors were made of vanadium permendur. This alloy is easily magnetized ($B = 2.3$ T at $H = 480$ A/cm). The neutron guide was placed as a whole into an electromagnet, 5 m long. The polarization of neutrons, averaged over the spectrum, was 80 % at a transmission of 60 %. The mean flux density of polarized neutrons at the guide exit was 3×10^6 neutrons/(cm^{2}s). The same neutron guide installed at the ILL reactor had the flux density at the guide exit of 3×10^8 neutrons/(cm^{2}s) [39].

The polarizing neutron guide of the Kurchatov Institute of Atomic Energy (IAE), installed at the IRT-M reactor [40], is also worth mentioning. A beryllium-water trap producing a peak in thermal neutron density was mounted at the center of the reactor core. A vertical channel traversing the entire water shielding of the reactor begins at this trap. A two-beam cobalt mirror poralizer with a total surface area of 2200 cm^2 was placed in the middle of the channel at a depth of 3 m underwater. A longitudinal field magnetizing the cobalt mirrors was produced by a solenoid wound on the channel. The polarization of the neutron beam was 75 % at the flux of 1.5×10^9 neutrons/s.

Drabkin and coworkers at LINP have developed and employed multilayered polarizing mirrors for polarizing neutron guides; the mirrors were fabricated as polished glass plates with a vacuum-deposited layer of 50 % Co–50 % Fe alloy and an underlying layer of 85 % Ti–15 % Gd alloy [41]. The composition of the ferromagnetic coating was so adjusted that the amplitudes b_N and b_m were almost equal, thereby increasing the neutron beam polarization. The underlying absorbing titanium-gadolinium layer separating the polished glass and the ferromagnetic coating suppressed the reflection from the glass surface of neutrons with spins antiparallel to the field. This structure resulted in a substantial extension of high polarization values to longer-wavelength neutrons (up to $\lambda = 0.7$ nm).

The first polarizing neutron guide at LINP, composed of such mirrors, was 1.47 m long, 1.6 mm wide, and 30 mm high; it had a curvature radius of 130 m and the characteristic angle of 5 mrad. The characteristic neutron wavelength was $\lambda^* = 0.27$ nm. The neutron guide was composed of seven mirror sections, 21 cm long each, and was evacuated to a pressure of 1.33 Pa. It was inserted into the gap between the poles of permanent magnets with a field strength of 400 A/cm. The tests produced a beam with the flux density of 1.7×10^7 neutrons/(cm^{2}s) and a beam polarization, averaged over the spectrum, of 0.97.

Similar mirror sections were used at LINP to fabricate a longer polarizing neutron guide (Fig. 1.6) [42]. Its length reached 5 m; channel width and height

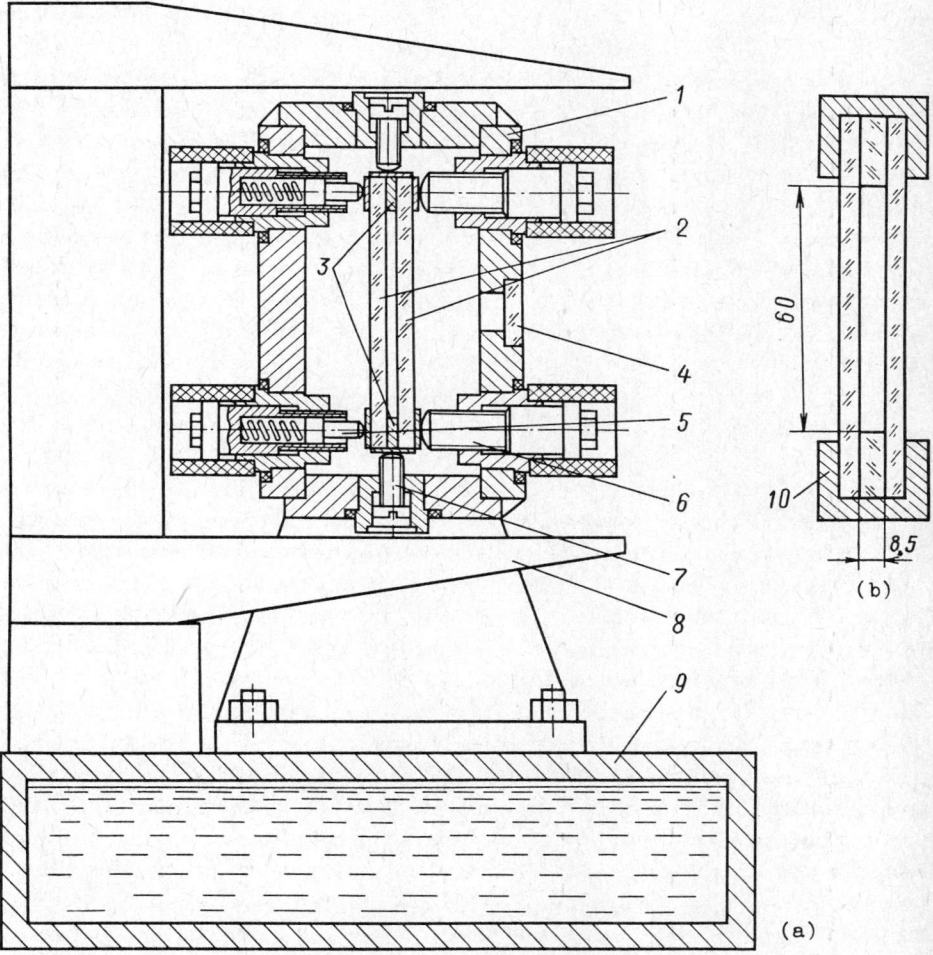

Fig. 1.6. The LINP polarizing neutron guide [42]: (a) neutron guide cross section, (b) cross section of the mirror channel; *(1)* vacuum housing, *(2)* polarizing mirrors, *(3)* spacers, *(4)* inspection ports, *(5)* spring supports, *(6.7)* adjustment screws, *(8)* poles of permanent magnets, *(9)* main girder, *(10)* clips

were 8.5 and 60 mm, respectively, at the characteristic angle of 7.4 mrad and characteristic wavelength of 0.4 nm. The upper and lower walls of the neutron guide channel were made of polished glass coated with a platinum layer. The neutrons transported by this guide peaked at $\lambda = 0.22$ nm; the beam had a flux density of 1.5×10^7 neutrons/(cm²s) and the mean polarization was 0.96.

An optical theory of multiple interference in multi-layer systems was developed in [43], and two-layer mirrors consisting of a 50 % Fe-50 % Co alloy layer (of various thicknesses) and an antireflection 74 % Ti-26 % Gd alloy layer were tested. The results obtained in [43] were similar to those reported in [41, 42].

Curved multislit polarizing neutron guides were developed and analyzed in [44]. Thin (80 µm) plastic foils made of polymethyl pentane with the empirical formula CH_2, coated in a vacuum on both sides by 150 nm layers of 60 % Co-40 % Fe alloy, were used as substrate. Owing to a large cross section of noncoherent scattering in the plastic foil, the substrate did not reflect neutrons with spins oriented antiparallel to the field. Plastic foils were glued to steel frames clamped between pairs of steel plates with grooves of the required curvature radius. This neutron guide had the following characteristics: channel width, 0.39 mm; curvature radius, 6.5 m; length, 18 cm; number of channels, 85; characteristic angle, 11 mrad; characteristic wavelength, 0.652 nm; beam cross section dimensions, 40×50 mm. The polarization of the beam at the exit of the neutron guide was 0.94.

The advantage of this system is its small size and mass (15 kg); its disadvantage is a considerable thickness of the mirrors (21 % of the channel width), resulting in a partial loss of the beam entering the neutron guide.

1.3.4 Multilayer Polarizing Monochromators

Multilayer mirrors were suggested as monochromators and polarizers of neutrons in [45, 46]. They function like interference mirrors in optics, and consist of alternating parallel layers of two substances of equal thickness. If one substance is ferromagnetic (M) and the other is nonmagnetic (V), and if the refractive index for neutrons with spins aligned parallel to the field in the ferromagnetic layer, n_{M+}, is made equal to the refractive index for neutrons in the nonmagnetic layer, $n_{V+} = n_{V-}$, this multilayer structure has a common reflection coefficient for neutrons with spins in parallel orientation; i.e., it functions as a single layer. If, in addition, we choose a ferromagnetic material such that $b_N = -b_m$; i.e., if we take a substance with negative nuclear scattering length equal in magnitude to the magnetic scattering length with the opposite sign, then $n_{M+} = n_{V+} = n_{V-} = 1$ and $n_{M-} > 1$. Neutrons with spins in parallel-to-field orientation (marked with the plus index) will not be reflected by the filter, while neutrons in the antiparallel-to-field orientation (marked with the minus index) undergo the Bragg reflection on the periodic structure, i.e., are reflected within a certain range of angles and wavelengths. In this mode, the reflected neutrons are polarized.

Such a multilayer monochromating polarizer was implemented at LINP [47]. Mirrors for the polarizer had alternating vanadium and Permalloy (^{62}Ni (66 %) and ^{54}Fe (34 %)) thermally deposited layers on a polished, 210×80×5 mm glass plate substrate. Up to 35 layers, 10 nm thick each, were deposited. The experimental results confirmed that the filter was highly reflective and polarizing: the maximum reflection coefficient was 75 % and polarization was 90–95 % at the reflection peak, at a neutron wavelength of 0.4 nm.

1.3.5 Polarizing Supermirrors

If the degree of monochromaticity of neutrons reflected by multilayer mirrors is deliberately reduced by slowly changing the period into the multilayered structure, an "all-wave" mirror can be obtained, reflecting the incident neutrons at any angle of incidence, in a sufficiently wide wavelength range.

This idea was first suggested by *Turchin* [48], and later independently by *Mezei* [49] who was able to realize it [50]. Such devices are called "supermirrors". Seventy-five alternating iron and silver layers, 7–40 nm thick, were vacuum-deposited on glass. Supermirrors assembled into a multislit Soller-type collimator yielded high polarization (97–99 %) in relatively weak magnetic fields ($H = 160$–320 A/cm).

A comparison of reflection coefficients and polarizations, as functions of the parameter θ/λ, is given for different mirrors in Figs. 1.7 and 1.8 [51]. The advantages of supermirrors are evident. Their shortcomings are the obvious complexity of fabrication and sharply decreasing degree of polarization at small glancing angles.

1.3.6 Systems Generating Polarized Neutron Beams

Table 1.1 lists the parameters of devices designed to produce polarized beams of nonmonochromatic neutrons. The list primarily covers systems intended for fundamental research which is the subject of the chapters that follow (for details, see [4] and the original papers).

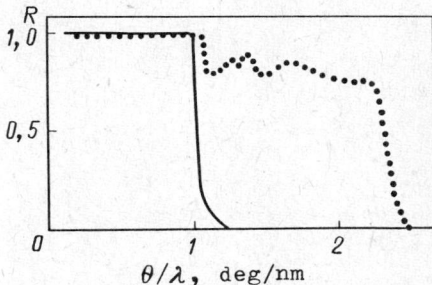

Fig. 1.7. Comparison of reflection coefficients R of ordinary 50 % cobalt – 50 % iron permendur mirrors (*solid curve*) and Mezei supermirrors (*dotted curve*) [51]

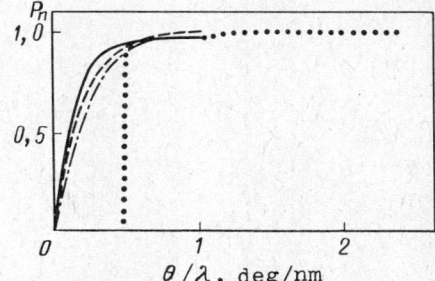

Fig. 1.8. Comparison of different totally reflecting polarizing mirrors by polarization P_n : (——) Drabkin's mirrors [41]; (- - -) mirrors on plasic substrate [44]; (-·-·-) Hamelin's mirrors [43]; (···) Mezei supermirrors [50]

Table 1.1. Characteristics of some apparatuses for producing polarized neutron beams

No.	Polarizer	Length [cm]	H [A/cm]	λ_{av} [nm]	Beam size [mm]	P_n [%]	Flux J [neutr/s]	Power of reactor [MW]	Ref.
1	Electrolytically depostied Co mirror	150	240	>0.1	10×100	90	$3 \cdot 10^7$	2.5	[52]
2	Multislit reflecting collimator composed of permendur mirrors	200	1200	>0.14	10×300	87	$3 \cdot 10^8$	5	[53]
3	Multislit focusing collimator composed of Co mirrors	105	240	>0.1	8×150	86±3	$4 \cdot 10^7$	2.5	[16]
4	Assembly of two Co mirrors	–	–	0.2	37×141	75±2	$4 \cdot 10^8$	10	[35]
5	Neutron guide made of permendur mirrors	500	680	0.6	5×57	80	$8.6 \cdot 10^6$	4	[38]
6	Multislit reflecting collimator composed of permendur mirrors	150	1200	0.2	20×100	90±5	$3 \cdot 10^8$	45	[36]
7	Assembly of two Co mirrors	160	320	>0.1	20×140	75	$1.5 \cdot 10^9$	7	[40]
8	Neutron guide with a CoFe layer and antireflecting TiGd sublayer	150	400	0.2	1.6×30	97	$8.2 \cdot 10^6$	16	[41]
9	Neutron guide composed of permendur mirrors	500	680	0.5	5×57	70±7	$3 \cdot 10^8$	57	[39]
10	Polarized proton target	2	$12 \cdot 10^3$	10^{-2}–10^5 eV	50×60	50	$3 \cdot 10^6$	0.02	[54]
11	Multislit neutron guide with FeAg supermirrors	60	200	–	–	98	–	–	[50]
12	Multislit neutron guide with CoFe layers on plastic foil	18	1120	0.65	40×50	94	–	–	[44]
13	Multislit neutron guide with CoFe layers on plastic foil	25	1120	0.56	30×50	90	$3 \cdot 10^9$	57	[55]
14	Multislit neutron guide with FeAg supermirrors	32	–	0.6	30×50	95–97	$4 \cdot 10^9$	57	[56]
15	Neutron guide with a CoFe layer and antireflecting TiGd sublayer	504	400	0.22	8.5×60	>96	$7.6 \cdot 10^7$	15	[42]
16	Multislit reflecting collimator composed of permendur mirrors	200	1600	–	30×80	>95	$6 \cdot 10^8$	44	[57]
17	Multislit neutron guide with CoTi supermirrors and antireflecting TiGd sublayer	300	–	–	15×50	93	$4 \cdot 10^8$	57	[56]

2. Precession of Magnetic Moments of Polarized Neutrons in a Magnetic Field

The precession of magnetic moments of polarized neutrons in a magnetic field is widely used in magnetic moment measurements and in the search for the electric dipole moment of the neutron. Knowledge of the magnetic dipole moments of the neutron and the proton – the main constituents of all nuclei – gives insight into the nature and structure of elementary particles [58].

Dirac's theory predicted the proton magnetic moment $\mu_p = \mu_N$, where $\mu_N = e\hbar/2m_p c \approx 5.05 \cdot 10^{-27}$ J/T is the nuclear magneton (m_p is the proton mass), and the neutron magnetic moment $\mu_n = 0$. However, Frisch and Stern [59] were able to show in 1933 that the proton magnetic moment exceeds μ_N by a factor of about 2.5, and the results of measuring the deuteron magnetic moment led to a hypothesis that the neutron has a negative magnetic moment of about $2\mu_N$. The first experiments on magnetic scattering of neutrons [60] already proved that neutrons have magnetic moments. The same experiments generated the first polarized neutron beams by using the magnetic interaction between neutrons and the atoms of ferromagnetic substances. Frish et al. [61] were the first to use polarized neutron beams for measuring the neutron magnetic moment. They obtained a qualitative result confirming that the neutron magnetic moment is negative, and roughly equals $2\mu_N$.

Quantitative measurements of the neutron magnetic moment became possible as a result of combining the techniques of neutron beam polarization and resonance depolarization of polarized beams, suggested by Rabi et al. [62]. The fundamentals of this method are presented in Sect. 2.1.

Experiments to find the dipole moment of the neutron were initiated by Purcell and Ramsey as early as 1950 [63]. An electric dipole moment of the neutron is implied by the symmetry properties of elementary particles and by quantum-mechanical conservation laws that are discussed in Chap. 6.

It should be recalled that in all interactions that conserve the spatial (P) parity, the electric dipole moment of an elementary particle, including the neutron, must be zero. However, violating the P parity would not be sufficient. The necessary condition for a nonzero electric dipole moment is, as shown by Landau [64], the simultaneous violation of both the spatial and the combined (CP) parities. Hence, a solid theoretical basis for searching for the neutron electric dipole moment appeared only after the discovery of the decay $(K_2^0 \to \pi^+ + \pi^-)$ of neutral K mesons, which violates the CP parity [65].

The neutron is an extremely attractive object in the search for the electric dipole moment because this particle participates in all four fundamental interactions: the strong, electromagnetic, weak, and gravitational ones. The properties of all fundamental interactions with respect to descrete transformations are, therefore, studied simultaneously. The neutron is electrically neutral, and can thus be subjected to strong electric fields; it will be shown later in the book that this fact is extremely important in the search for the electric dipole moment. Neutron beams generated by nuclear reactors are easily polarized and have a high intensity. Finally, the average lifetime of neutrons is comparatively long (about 900 s), so that neutrons interact comparatively easily with the experimental system.

2.1 Fundamentals of the Magnetic Resonance Depolarization Technique

Measurements of both the magnetic dipole and the electric dipole moments of the neutron make use of spin precession, and hence, of the precession of neutron magnetic moments in a uniform permanent magnetic field H_0, as described by (1.1). The precession frequency, the so-called Larmor frequency, is given by the formula

$$\omega_L = \gamma_n H_0 . \tag{2.1}$$

Switching to quantum-mechanical terms, we can say that the neutron magnetic moment μ_n is allowed to have one of two orientations in a magnetic field H_0 : parallel to the field and antiparallel to it. The energies of these states are given by the expression

$$E = -\boldsymbol{\mu}_n \boldsymbol{H}_0 \tag{2.2}$$

and differ by a quantity

$$E = 2\mu_n H_0 = \hbar \omega_L .$$

This difference can be measured if an oscillating magnetic field $H_1 \cos(\omega t)$ at a frequency ω equal to the Larmor frequency ω_L is applied in the plane perpendicular to H_0. Under these conditions, the energy of the oscillating field, $\hbar\omega_L$, becomes equal to the energy difference ΔE and produces the resonance transitions between the two states of the neutron. The energy ΔE is taken from (or passed on to) the oscillating field. If the neutron beam entering the region with the field H_0 is polarized, then the number of transitions from the states with a larger population is greater than that from the states with a smaller one. As a result, the populations tend to equalize and the beam polarization diminishes. At $\omega = \omega_L$, the resonance depolarization of the neutron beam occurs. This resonance depolarization technique, suggested by *Rabi* [62], is described in detail in [66].

The probability for the neutron spin, placed in a field H_0 and subjected for a time t to an oscillating field $H_1 \cos(\omega t)$ at right angles to H_0, to reverse its orientation with respect to the field H_0 is given by (1.12). The flight time of neutrons through the oscillating field is sufficiently long in the Rabi method (a long coil is employed).

2.2 Measurements of the Neutron Magnetic Moment

2.2.1 Survey of Experiments

Let us trace the gradual elaboration of experimental techniques for measuring the neutron magnetic moment. Quite a few achievements resulted from the studies which were stimulated by the need for higher precision in measuring this important constant. The Rabi resonance method was first applied to neutrons by *Alvarez* and *Bloch* [11]. The greatest difficulties encountered by these methods are associated with measuring the absolute value of the frequency ω and the field strength H_0 at the resonance. The source of neutrons in [11] was a cyclotron (deuterons were bombarding a beryllium target), so that it proved useful to work at resonance

$$\omega_p = eH_p/m_p c ,$$

for protons revolving in the cyclotron, where ω_p is the cyclotron resonance frequency, and H_p is the resonance field strength in the cyclotron. A comparison of ω with ω_p and of H_0 with H_p yields the quantity

$$\gamma_n = (\omega/\omega_p) H_p/H_0 . \tag{2.3}$$

Having measured γ_n, Alvarez and Bloch determined that the neutron magnetic moment was

$$|\mu_n| = (1.935 \pm 0.020)\,\mu_N . \tag{2.4}$$

The method was further refined after the discovery of nuclear magnetic resonance (NMR) [67]. In fact, this discovery was stimulated to a large extent by experiments aimed at measuring the neutron magnetic moment. The neutron magnetic moment was measured relative to the proton magnetic moment by comparing the frequency of the oscillating field of the resonance coil, which depolarized the polarized neutron beam in a uniform permanent magnetic field, with the proton resonance frequency in the same field. This method improved the precision of measuring the neutron magnetic moment [12], reaching the accuracy with which the proton magnetic moment was known at that time:

$$|\mu_n| = (1.9103 \pm 0.0012)\,\mu_N . \tag{2.5}$$

Bloch et al. [13] achieved still better accuracy in their measurements of the neutron magnetic moment, using the same technique as in [12]. The neutron magnetic moment they reported in [13] was

$$\mu_n = -(1.91307 \pm 0.00006)\, \mu_N\ . \tag{2.6}$$

In 1950, *Ramsey* [68] suggested another refinement of the technique. He replaced one long coil generating the oscillating magnetic field with two coils which produced this field over short segments at the beginning and end of the permanent field region. The method of two separated rf fields has considerable advantages over the Rabi method. Since creating a high-strength, perfectly uniform magnetic field is virtually unfeasible, the resonance frequencies at different parts of the magnet are unequal, resulting in a broadening of resonance peaks. The Ramsey method eliminates the effects of magnetic field nonuniformity on the resonance width and produces a narrower resonance curve. The details of this technique can be found in [69]. Ramsey also showed that it was useful to introduce a phase shift between the two oscillating fields. The intensity curve was of the dispersion type, passing through zero at the resonance frequency ω_L.

Precise measurements of H_0 were carried out at Brookhaven National Laboratory (USA) using NMR techniques, while the resonance frequency ω_L was measured by separated rf fields [70]. A proton-containing specimen (distilled water probe) was placed together with the generating and detecting coils into the field H_0 in which the resonance depolarization of neutrons occured. The proton resonance frequency was measured by displacing the water probe along the magnet axis between the rf coils at a step of 0.5 cm, and the resonance frequencies were then averaged. The obtained value of the neutron magnetic moment was

$$|\mu_n| = (1.913148 \pm 0.000066)\, \mu_N\ . \tag{2.7}$$

The highest accuracy in measuring the neutron magnetic moment was achieved by *Ramsey* et al. [71] at the ILL with a system designed to look for the neutron electric dipole moment [5]. In the next subsection we give a more detailed description of this system (for details on the search for the neutrons electric dipole moment in this work, see Sect. 2.3).

2.2.2 The Magnetic Resonance Neutron Spectrometer at Grenoble

A neutron beam from the high flux reactor was directed through a neutron guide at a magnetized iron mirror polarizer. The polarized neutron beam was sent to a Pyrex glass tube, 1.1 cm in diameter, used as a neutron guide and passing through an electromagnet producing a homogeneous magnetic field $H_0 = 14.4$ A/cm. Two rf coils, 3 cm long each, producing the oscillating magnetic field, were placed between the magnet poles at the beginning and the end of the magnetic field region, 60 cm apart. The rf coil oscillation phases were shifted by ±90°. Beam polarization was analyzed by reflecting the beam off the second magnetized mirror. Neutron detectors were scintillators of ^{6}Li-loaded glass glued directly onto photomultiplier tube windows. It was, therefore, possible to record very high neutron fluxes (up to 5×10^6 n/s). The neutron resonance width was 124 Hz at the resonant frequency of 52.3 kHz.

2.2.3 The Value of the Neutron Magnetic Moment

The ratio of the neutron to proton magnetic moments, after certain small corrections, was found to be

$$\mu_n/\mu_p = -(0.68497945\pm0.00000017) \; . \tag{2.8}$$

Using the data [72] on the proton magnetic moment, the authors of [71] obtained

$$\mu_n = -(1.91304184\pm0.00000088)\,\mu_N \; . \tag{2.9}$$

The work of *Greene* et al. [71] represents one of the highest-precision experiments measuring a fundamental physical constant. The value of μ_n reported in [71] diverges from the result (2.7) in [70] but is in agreement with the data (2.6) in the earlier paper [13].

2.2.4 The Sign of the Neutron Magnetic Moment

The negative sign of the neutron magnetic moment was already established by the first experiments with polarized neutrons [61]. The same conclusion was reached by *Bloch* et al. [13], but their arguments were not convincing because of the smallness of the observed effects.

Rogers and *Staub* [73] directly compared the signs of the proton and neutron magnetic moments, and determined these signs. They replaced the oscillating magnetic field used in [13] by a revolving field. They used crossed coils at right angles to each other, supplied with rf signals differing in phase by 90°. The resonance was observed both for protons and neutrons only in a certain direction of revolution of the fields, pointing in the direction of the Larmor precession. The direction of revolution for protons was found to be the opposite of that for neutrons. The revolution directions proved that the proton magnetic moment is positive while the neutron magnetic moment is negative.

The negative sign of the neutron magnetic moment was recently confirmed in [74]. The authors made use of the resonance reversal of the polarization of neutrons moving through a spatially periodical magnetic field.

2.2.5 Comparison with the Theory

As we have mentioned, the experimental results contradict the predictions of Dirac's theory. An attempt is made in the meson theory of nuclear forces to explain this discrepancy by the presence around nucleons of a cloud of virtual mesons whose motion produces a circular current, generating both the anomalous magnetic moment of the neutron and the anomalous part of the proton magnetic moment.

The nature of the neutron magnetic moment was discussed in detail in [34]. The experiment [37] demonstrated the validity of Schwinger's model in which the neutron is treated as an element of current, not as a point-like dipole (the Bloch model).

The quark model of nucleons gives the ratio of the neutron to proton magnetic moments $\mu_n/\mu_p = -2/3$. The same result is given by the SU(6) symmetry model applied to the baryon octet (see [75]). The experiment confirms this prediction of the theory fairly well.

2.3 Search for the Electric Dipole Moment of the Neutron

2.3.1 Fundamentals of the Method

The precession of polarized neutrons in a permanent magnetic field, employed for measuring magnetic moments, is also used in the search for the neutron dipole electric moment d.

If the neutron possesses a nonzero electric dipole moment, then the only direction allowed for this moment is that of the neutron spin. In order to detect the electric dipole moment in a magnetic resonance spectrometer designed for magnetic moment measurement, a high-strength electric field E was applied parallel to the magnetic field. In this case the energy of interaction with the electric field, dE, dependent on the mutual orientation of the vectors d and E, is added to the energy (2.2) of the neutron interaction with the magnetic field. Consequently, if the direction of the magnetic field, and hence, that of the neutron spin, remain unchanged but the direction of the electric field is reversed, a term $-2d\Delta E$ appears in the difference between the energies of the neutron-field interaction; this term shifts the resonance frequency by

$$\Delta \nu = -2d\Delta E/h = -4dE/h , \qquad (2.10)$$

where $\Delta E = 2E$ is the algebraic difference between electric field strengths in the applied fields. The minus sign appears because the neutron magnetic moment and the electric dipole moment (provided it exists and has a plus sign) are oriented counter to each other. At a constant frequency of the oscillating magnetic field, a change in the resonant frequency results in shifting the resonance curve, i.e., in changing the counting rate in the neutron detector by

$$\Delta N = N_+ - N_- = -(dN/d\nu)\Delta \nu , \qquad (2.11)$$

where $dN/d\nu$ is the slope of the resonance curve, and $N_{+(-)}$ is the counting rate of the detector for the parallel (antiparallel) orientation of the electric field with respect to the magnetic field.

The neutron electric dipole moment follows from (2.10):

$$d = -h\Delta\nu/4E \ . \tag{2.12}$$

The sensitivity of the experiment increases as the electric field strength E and the slope $dN/d\nu$ of the resonance curve increase. Nowhere was a shift in the resonant frequency detected.

The first experiments were carried out by *Smith* et al. [76] at Oak Ridge National Laboratory (USA) and gave for the upper bound on the electric dipole moment the value $d < 5 \times 10^{-20}$ e×cm (the electric dipole moment is measured in units of the electron charge e times the unit length, cm). The sensitivity of the system was then gradually improved. Many laboratories engaged in the search for the neutron electric dipole moment, but the most significant contribution was made by *Ramsey* and his coworkers. The evolution of Ramsey's instrument is described in [75].

The experiment that achieved the highest sensitivity in the search for the electric dipole moment of the neutron using the magnetic resonance depolarizatioin technique, was completed by Ramsey and his coworkers in 1976 at the ILL reactor at Grenoble [5]. Let us discuss this experiment in more detail.

2.3.2 The Experiment on Measuring the Electric Dipole Moment of the Neutron

The magnetic-resonance neutron spectrometer used in this study is the one described in Sect. 2.2.2, subsequent to modifications dictated by electric dipole moment measurements (Fig. 2.1).

The homogeneous magnetic field of strength $H_0 = 13.6$ A/cm was produced in the pole gap 9 cm wide by permanent magnets made of Alnico alloy. The electric field strength $E = 100$ kV/cm, parallel to the magnetic field, was produced in a gap 1 cm wide along a path 180 cm long by an electrostatic capacitor. Two rf coils working at a frequency of 50 Hz were placed 200 cm apart. The spectrometer was mounted on a rotating turntable and could be rotated 180° around the vertical axis together with the polarizer and the analyzer. This last feature was required for the elimination of the main systematic error arising when the magnetic and electric fields are not parallel (this factor will be discussed later in this section).

Let us discuss some of the requirements for spectrometers used in this search for the neutron electric dipole moment by the magnetic resonance depolarization technique, and see how these requirements are met in the experiment [77].

The theory of magnetic-resonance spectrometers with two separated coils [69] shows that each neutron passing through the spectrometer is subjected to the average static magnetic field between these coils, and that it is this average field which determines the resonance frequency. If all neutrons moved along the same trajectory, the resonance frequency for them would be identical.

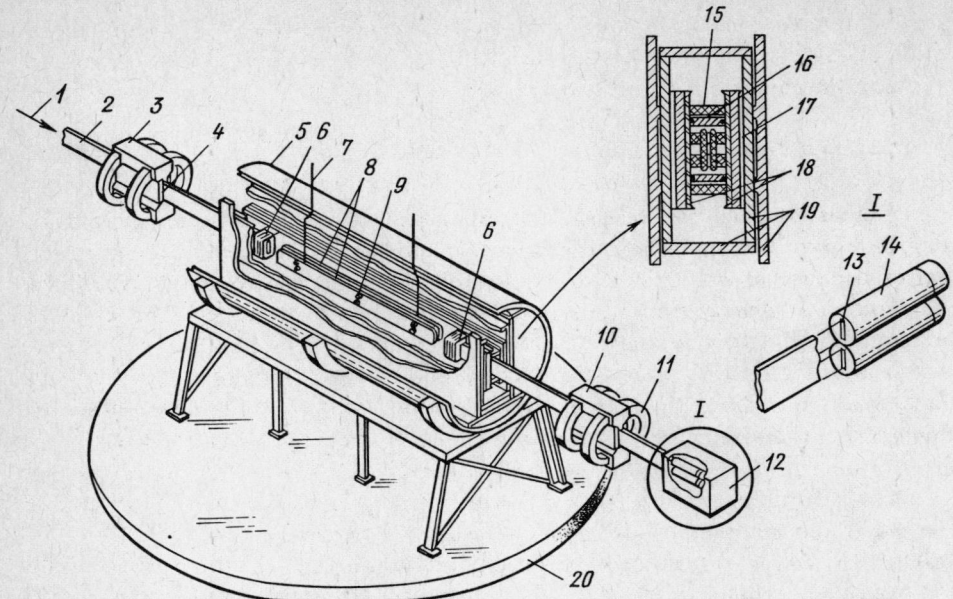

Fig. 2.1. Experiment measuring the electric dipole moment of the neutron [5]: *(1)* neutron beam, *(2)* neutron guide tube, *(3)* spin polarizer, *(4)* polished iron mirrors, *(5)* double magnetic Molypermalloy shield, *(6)* rf coil, *(7)* electric field source, *(8)* electrostatic plates, *(9)* insulators, *(10)* spin analyzer, *(11)* Alnico magnets, *(12)* detector, *(13)* scintillator strips, *(14)* photomultipliers, *(15)* quartz spacers, *(16)* vacuum chamber walls, *(17)* Alnico permanent magnets, *(18)* Milypermalloy pole pieces, *(19)* soft iron, *(20)* rotating turntable

However, beam dimensions being 10×1 cm, it becomes necessary to provide a homogeneous magnetic field across the whole region occupied by the beam. This is necessary in order that the resonance maximum for some neutrons not overlap the minimum for other neutrons. The average velocity of neutrons in the beam is $v \approx 150$ m/s, and the coils are spaced at $l \approx 2$ m. The halfwidth of the magnetic resonance in frequency units, $\Delta\nu/2$, can then be found from the uncertainty relation. The uncertainty in energy is

$$\Delta E \geq h \Delta \nu = h/\Delta t \; ,$$

where $\Delta t = l/v$ is the time a neutron spends in the spectrometer. Hence, $\Delta\nu/2 \approx 40$ Hz. For this reason, the variations of the mean magnetic field strength for neutrons moving along different trajectories must be limited:

$$\Delta H < h \Delta \nu / 2 \mu_n \approx 10^{-2} \text{ A/cm} \; . \tag{2.13}$$

The temporal stability of the magnetic field must be such that the resulting error be much less than that expected for the Poisson distribution of detector counts. This error is equivalent to a shift $\Delta\nu$ in the resonant frequency. Estimates show that over a measurement time $\Delta T = 200$ s, the sensitivity of this system reaches $3\times10^{-3}\,e\times\mathrm{cm}$, which corresponds to a resonant frequency shift of $\Delta\nu\approx3\times10^{-3}$ Hz. At the resonant frequency $\nu = 50\,\mathrm{kHz}$, the required temporal stability of magnetic field must be

$$\Delta H/(H\Delta T) = \Delta\nu/(\nu\Delta T) = 6\times10^{-8}/200\,\mathrm{s}^{-1}\ . \tag{2.14}$$

Let us consider one of the most important requirements, namely, a high degree of parallellism of the static magnetic and electric fields. This requirement arises because of a spurious electric dipole moment due to the motion of a neutron at a velocity v in an electric field of strength E. Indeed, the effective magnetic field acting on the neutron in the comoving reference frame is $H_\mathrm{eff} = -(1/c)v\times E$. The static magnetic field H_0 and the effective field form the total field of strength $H = H_0 + H_\mathrm{eff}$. If the vectors H_0 and E are at an angle θ to each other, the strength of the resultant field is

$$H = H_0 \pm H_\mathrm{eff}\sin\theta\ ,$$

where the plus and minus signs correspond to the opposite directions of the electric field. When the electric field is reversed, the strength of the magnetic field acting on the neutron changes by a quantity

$$\Delta H = 2(v/c)E\sin\theta\ ,$$

which shifts the resonant frequency by

$$\Delta\omega = \gamma_\mathrm{n}\Delta H\ .$$

This is equivalent to the neutron having a spurious electric dipole moment equal to

$$\mu_\mathrm{n}(v/c)\sin\theta \approx \mu_\mathrm{n}(v/c)\theta\ .$$

The effect of this moment is negligible if the inequality

$$\mu_\mathrm{n}(v/c)\theta \ll d$$

holds, giving a constraint on the angle θ :

$$\theta \ll dc/\mu_\mathrm{n}v\ . \tag{2.15}$$

This means that for neutrons moving at a velocity of $150\,\mathrm{m/s}$, at spectrometer sensitivity of about $10^{-24}\,e\times\mathrm{cm}$, the angle θ must be less than $0.01°$. On the average, the angle between E and H_0 must not exceed this value in the whole region occupied by the electric field.

The effect of the spurious electric dipole moment can be isolated experimentally by changing the sign of v, i.e., by rotating the whole spectrometer

180° around the vertical axis, and subtracting this effect from the experimental result.

Let us see now to what extent the constraints on the homogeneity and temporal stability of the magnetic field and on the parallelism of the magnetic and electric fields were met in the described system [5].

Magnetic Field Homogeneity ($\Delta H \approx 10^{-4}$ A/cm) was achieved by specially designing the poles, first of all by adjusting the pole size normally to the beam axis (30 cm for the beam height of 10 cm and pole gap width of 9 cm). The inner part of the poles was made of Molypermalloy (79 % Ni, 17 % Fe, and 4 % Mo) which ensured a high degree of magnetic equipotentiality of the poles, owing to the high magnetic permeability of the alloy. Triangular shims were used to eliminate edge effects. Special measures were taken to keep the poles rigorously parallel. First of all, the faces of the pole tips were made parallel by machining to within 7.5 μm. The gap width between the pole tips was maintained constant by a number of calibrated quartz spacers. The inner surface of the poles formed the walls of the vacuum chamber through which the neutron beam passed. The middle part of the poles was made of Armco steel, and was separated from the inner part by a narrow evacuated gap introduced for equalizing the pressure on both sides of the inner poles. To minimize the distortion of the magnetic field's lines of force, only materials with low magnetic susceptibility were used inside the pole gap.

Stability of the Magnetic Field was maintained at a level of $5 \times 10^{-8}/200\,\text{s}^{-1}$ by making the outer part of permanent magnet poles of Alnico alloy whose magnetization is only weakly dependent on temperature. The field was also stabilized by using a magnetic shield consisting of two concentric cylindrical Molypermalloy layers, 2.5 mm thick, and planar layers coating the end faces. The resulting shielding factor $k = H_{\text{out}}/H_{\text{in}}$ reached 800.

Parallelism of the Magnetic and Electric Fields was achieved by precision machining of the Molypermalloy pole surfaces, of the plates of the electrostatic capacitor, and of the supporting insulators. It was found that the angle $\theta < 0.006°$.

Measurement Procedure. Figure 2.2 shows neutron resonance curves for the phase difference between rf coil fields $\Delta\varphi = \pm 90°$. The resonant frequency was determined to be the one corresponding to the point of intersection of the two curves to the right of the vertical dash-dot line. The measurement cycle consisted in changing the phase difference betwen coils once a second and reversing the electric field every 200 seconds ($\Delta T = 200\,\text{s}$).

Changing the phase difference from +90° to -90° replaced one resonance curve by another that was antisymmetric to it with respect to the true resonant frequency, so that the true resonant frequency was found from the change in counting rates at the phase differences of +90° and -90°. In effect, the difference

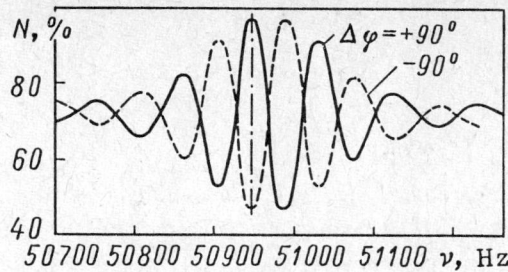

Fig. 2.2. Neutron resonance curves [5]. The ordinate is the counting rate N as a percentage of the rate with no rf, plotted as a function of the frequency ν of the oscillating magnetic field produced by rf coils

between counting rates at a fixed coil frequency (as close as possible to the true Larmor frequency) for the phase differences of $+90°$ and $-90°$ was divided by the slope of the resonance curve.

This procedure yielded 100 values of resonant frequency, every 200 seconds (Fig. 2.3). Straight lines were drawn through these points by the least squares fit, yielding the average values of the resonant frequencies $\bar\nu_+$ and $\bar\nu_-$. The frequency difference $\Delta\nu$ was found by extrapolating the lines drawn through the average values of the resonant frequencies to the midpoint in time between the moments of electric field reversals.

The advantage of this measurement procedure was that it averaged out, in the first approximation, the short-period (about 1 s) instrument noise and instabilities. The contribution of the noise and instabilities with periods close to 200 s was also small. The results contained, together with true or induced signals appearing in sync with electric field reversals, only long-period trends connected with daily variations of temperature and neutron flux in the reactor cycle.

The measurement cycle was repeated several hundred times per day. In order to eliminate the non-parallelism of the E and H_0 fields, the spectrometer was rotated $180°$ around the vertical axis once very 24 hours.

Other Systematic Errors. One of the significant sources of systematic error in the experiment was the change in the magnetic field correlated with

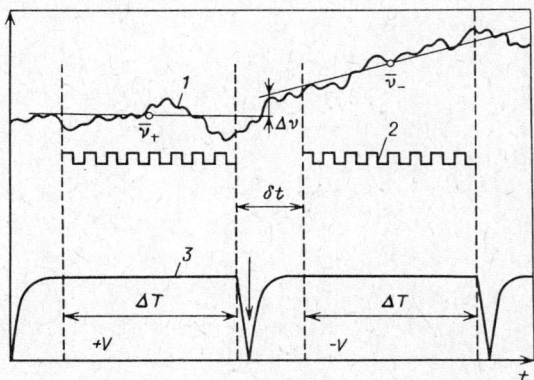

Fig. 2.3. Basic measurement cycle [5]. *(1)* instantaneous resonant frequency with high- and low-frequency components (noise, instability, temperature drift), *(2)* phase difference between the oscillations of two rf coils ($\Delta\varphi = +90°$ and $\Delta\varphi = -90°$), *(3)* high voltage applied to electrostatic capacitor plates; ΔT is the measurement time, δt is the switching interval; the average resonant frequencies $\bar\nu_+$ and $\bar\nu_-$, and frequency difference $\Delta\nu$ are shown. The time scale is distorted. The arrow indicates the moment of polarity reversal

the reversal of electric field polarity. Estimates showed that a spectrometer sensitivity of $d\approx 10^{-24}$ e×cm could be achieved if the magnetic field variation accompanying the electric field reversal did not exceed 10^{-7} A/cm. This constraint imposes a 10^{-7} A upper bound on the current passing through the spectrometer and reversing its direction in sync with electric field polarity reversal. Earlier experiments by the same research group detected a current of the order of several microamperes between the spectrometer and the recording instruments. Complete electric insulation of the spectrometer reduced this current to 10^{-12} A, eliminating this systematic error.

Another source of systematic error was sparking in the electric field. Spark breakdown caused local magnetization of the Molypermalloy poles, and hence, shifted the resonant frequency by 10–50 mHz. Sparking occured in spite of a helium buffer gas filling of the whole inner space of the spectrometer at a pressure of 10^{-5} hPa. The effects of sparking on the results were eliminated by neglecting the results recorded immediately before and after each spark breakdown.

Systematic effects connected with the displacement of the plungers of the device reversing the direction of the electric field, overhead crane motions, compression of quartz spacers by the electric field, and some other effects were evaluated and their sources partially eliminated.

Data Acquisition. The results were stored on magnetic tape every 200 s. The data included the average values of the resonance frequencies, the slopes and the intercept of the least-squares fit of the 100 phase-reversed pairs of true resonance frequencies, the root mean square deviation of instantaneous frequencies from their mean values, the ratio of this deviation to the statistical error, the value of the electric field, the magnetometer reading, the number of spark breakdowns, and the thermometer reading.

Before the final processing stage, the data were corrected for the daily temperature drift of the setup, and averaged over 24 hours. Each day's results were collected into three groups representing the results of two or three months of observation. In each group systematic errors were subtracted from the obtained average values.

Some measurements were conducted with a zero electric field. Their results were also subtracted from the results of measurements with a nonzero field.

2.3.3 The Result of Searching for the Electric Dipole Moment of the Neutron

The experiment yielded the following value of the neutron electric dipole moment [5]:

$$d = (0.4 \pm 1.5) \times 10^{-24} \text{ e} \times \text{cm} \ . \tag{2.16}$$

The authors of [5] interpreted this result as an upper bound on the neutron electric dipole moment, at the 90 % confidence level:

$$|d|<3\times10^{-24}\,\mathrm{e\times cm}\,. \tag{2.17}$$

This result is lower by four orders of magnitude than the limit established by their earlier experiment [76].

The possibilities of this method of measuring the electric dipole moment of the neutron are completely exhausted [5], mostly because it is not feasible to improve the degree of parallelism of the magnetic and electric fields (see constraint (2.15)).

2.3.4 Other Experimental Methods in the Search for the Electric Dipole Moment of the Neutron

Among the methods of searching for the electric dipole moment of the neutron, different from that of [5], are the crystal diffraction method and the method of magnetic resonance depolarization employing ultracold neutron beams.

So far, the crystal diffration method has not yielded results comparable in accuracy with those obtained by magnetic resonance depolarization (see Sect. 2.1). The only paper on measuring the neutron electric dipole moment by the crystal diffraction technique reported the upper bound $d<5\times10^{-22}\,\mathrm{e\times cm}$ [78], which is higher, by two orders of magnitude, than the result (2.17). Suggestions were made on how to improve the sensitivity of the crystal diffraction method; nevertheless, it is unlikely that the sensitivity of the magnetic resonance depolarization method will be surpassed here. Detailed information on the crystal diffraction method of searching for the neutron electric dipole moment and on the expected sensitivity can be found in [75].

Contrary to the crystal diffraction method, the method of magnetic resonance depolarization applied to ultracold neutron beams has already produced results whose accuracy exceeds that for thermal neutrons. The work on generating ultracold neutron beams was initiated in 1958 by *F.L. Shapiro* [79], who wanted to measure the neutron electric dipole moment. Indeed, the neutron-apparatus interaction time can be greater by several orders of magnitude for ultracold neutrons than for thermal neutrons.

Using ultracold neutrons instead of thermal neutrons also suppresses, by several orders of mangitude, the main source of systematic error plaguing the magnetic resonance depolarization techniques, namely, the nonparallelism of the E and H_0 fields, dependent on the neutron velocity.

However, the work with ultracold neutrons is beyond the scope of this book, so that we shall give only the most accurate results obtained by this technique. *Lobashev* et al. [80] at LINP reported the following upper bound on the electric dipole moment of the neutron, at the 90% confidence level:

$$|d|<6\times10^{-25}\,\mathrm{e\times cm}\,. \tag{2.18}$$

This value is less, by a factor of 5, than the upper bound (2.17) obtained by magnetic resonance depolarization of thermal neutrons.

2.3.5 Comparison with the Theory

Theoretical prediction of the value of d depends on which interaction is regarded as the source of violation of the CP invariance.

If the CP invariance is broken by the superweak interaction changing strangeness by $|\Delta S| = 2$, then d is immeasurably small, less than 10^{-38} e×cm. If there exists a strangeness-conserving superweak interaction ($\Delta S = 0$), then $d \approx 10^{-28}$ e×cm. All these theoretical predictions are much less than the experimentally established limit (2.18).

The largest predictions of d, not much smaller than the experimental upper bound (2.18), are implied by hypotheses based on the *Weinberg* model [81] in which the CP invariance is broken by the exchange of charged Higgs bosons introduced in the standard electroweak interaction model [82]. For more details, see Okun's monograph [58].

3. Interaction of Polarized Neutrons with Polarized Nuclei

The study of spin states of the levels of the compound nuclei, formed as a result of the interaction between nuclei and slow neutrons, is of considerable importance for clarifying the dependence of nuclear forces on spin. Thus, it is very interesting to find a correlation between the spin state of a level of a compound nucleus and other parameters of this level, such as the radiative and neutron widths, the spacing between the levels, and the spectrum of the capture gamma rays. It is also important to study the correlation of neutron strength functions $S_0 = \overline{\Gamma}_n^0/D$ with resonance spins $J_c = J_i \pm 1/2$ ($\overline{\Gamma}_n^0$ is the mean reduced neutron width, D is the mean distance between the levels, and J_i is the spin of the original nucleus). All these data are required to test the suggested models of nuclei.

Only a few methods make it possible to determine the spin states of the levels of a compound nucleus. Polarized neutron beams opened up new possibilities in this field. One such possibility is realized in the studies of the interaction between polarized neutrons and polarized nuclei.

3.1 Fundamentals of the Method

A system of polarized nuclei is usually characterized by the orientation parameters $f_k(J_i)$ defined by *Tolhoek* and *Cox* [83]. According to [83], the following three orientation parameters are the most important:

$$f_0 = 1 \; ; \quad f_1 = \left(\sum_m a_m m\right)/J_i \; ;$$

$$f_2 = \frac{1}{J_i^2}\left[\sum_m a_m m^2 - \frac{J_i(J_i+1)}{3}\right] \; , \qquad (3.1)$$

where a_m are the populations of magnetic sublevels with magnetic quantum numbers $J_z = m$. A system of nuclei is said to be polarized if at least one of the orientation parameters $f_k(J_i)$ with odd k has a nonzero value. By definition, the polarization of nuclei in a target is the quantity

$$P_N = (N_+ - N_-)/(N_+ + N_-) \; , \qquad (3.2)$$

where $N_{+(-)}$ is the number of nuclei with spins parallel (antiparallel) to a specified direction in space (this direction is usually fixed by the orientation of the magnetic field) per unit volume of the target. The polarization of nuclei equals the first orientation parameter: $P_N = f_1$.

The total cross section of the interaction between s-neutrons, that is, neutrons with zero orbital momentum, $l = 0$, and the nuclei of a polarized target is written in the form

$$\sigma = \sigma_0 + P_n P_N \sigma_p = \sigma_0(1 + \varrho P_n P_N) \;, \tag{3.3}$$

where σ_0 is the cross section for an unpolarized target and an unpolarized beam; P_n is the polarization of the neutron beam (usually P_n and P_N are of like signs if the directions of the predominant orientations of the neutron and nucleus spins coincide); σ_p is the polarization cross section; and $\varrho = \sigma_p/\sigma_0$. As a rule, the potential scattering cross section and the cross section of compound nucleus formation are separated within the total cross section of interaction between neutrons and medium and heavy nuclei. If a neutron and a target nucleus interact via a compound nucleus level with a definite spin state, the polarization cross section is [84]:

$$\sigma_p = \begin{cases} \sigma_p = \sigma_0 J_i/(J_i + 1) \;; & J_c = J_i + 1/2 \;, \\ \sigma_p = -\sigma_0 \;, & J_c = J_i - 1/2 \;, \end{cases} \tag{3.4}$$

where J_i is the spin of polarized nuclei, and J_c is the spin of compound nuclei.

In its turn, σ_0 can be written in the form

$$\sigma_0 = \frac{J_i + 1}{2J_i + 1}\sigma_+ + \frac{J_i}{2J_i + 1}\sigma_- \;, \tag{3.5}$$

where σ_+ and σ_- are interaction cross sections in the spin states $J_i + 1/2$ and $J_i - 1/2$, provided the coherent effects are negligible in the interaction of neutrons with target nuclei [34].

If the interaction of neutrons with target nuclei goes via both spin states of the levels of the compound nucleus, then

$$\sigma_p = [J_i/(2J_i + 1)](\sigma_+ - \sigma_-) \;. \tag{3.6}$$

Suppose that a polarized neutron beam with polarization P_n passes through a d-thick polarized target with nuclear polarization P_N, having N nuclei per unit volume. Then the target transmission, defined as the ratio of the neutron flux in the neutron detector to the neutron flux with the target removed, is given by the expression

$$T = T_0[\cosh(P_N N \sigma_p d) - P_n \sinh(P_N N \sigma_p d)] \;, \quad \text{where} \tag{3.7}$$

$$T_0 = \exp(-N \sigma_0 d)$$

is the transmission of the same target in the case of unpolarized neutrons and nuclei.

The replacement of the prallel orientation of the spins of neutrons and nuclei by the antiparallel orientation results in a relative change in transmission (provided $P_N N \sigma_p d \ll 1$)

$$\varepsilon = \frac{T_p - T_a}{T_p + T_a} = -P_n \tanh(P_N N \sigma_p d) \approx -P_n P_N N \sigma_p d \;, \tag{3.8}$$

where $T_{p(a)}$ is the target transmission for the parallel and antiparallel mutual orientations of the neutron and nuclear spins. As follows from the expressions (3.8) and (3.4), the sign of ε is dictated by the sign of the polarization cross section σ_p, and hence, we can distinguish between two cases:

$$J_c = J_i + 1/2 \quad \text{and} \quad J_c = J_i - 1/2 \;.$$

Of course, it is necessary to know the absolute directions of neutron and nuclear spins; they are rather easily found. The value of ε depends on what resonance is studied, but as a rule it is not less than 1.5%, so that it is not difficult to determine the sign of ε, and thus find the spin J_c of the compound nucleus. In addition to the data on the spins of the levels of compound nuclei, the results on the transmission of polarized neutrons through polarized nuclear targets yield the nuclear polarization P_N, the mean temperature of the nuclear target, and the hyperfine splitting constant.

3.2 Survey of Experiments

Rose [84] was the first to point out that the capture of polarized slow neutrons by polarized nuclei can be used as a basis for determining the spins of the levels of compound nuclei.

The first experiments were carried out by *Bernstein* et al. [85] in 1953. The method employed for the polarization of ^{55}Mn nuclei was the Gorter-Rose magnetic hyperfine splitting technique. The activity of the formed ^{56}Mn nuclei was studied as a function of mutual orientation of the neutron and nuclear spins. The results ($\varepsilon \approx 4\%$) proved that the cross section of the neutron capture by ^{55}Mn nuclei depends on the mutual orientation of their spins.

Later this group studied the polarization of neutrons produced when an initially nonpolarized beam passes through a polarized ^{149}Sm target [86]. The next stage was the transmission of 0.075 eV polarized neutrons by a metallic indium target as a function of the mutual orientation of the neutron and nuclear spins [87]. The ^{115}In nuclei were polarized by the external field method (the Simon method).

In the 1960s, a number of papers, using similar experimental methods, reported the successful determination of the spins of several levels of compound nuclei [88].

The method of measuring the interaction of polarized neutrons with polarized nuclei became more widespread after the polarized proton target was developed as a neutron polarizer. This system was first realized at the Neutron Physics Laboratory (LNP) at the Joint Institute of Nuclear Research (JINR) by *Shapiro* and coworkers [89–91] in 1963 (see [4]). The advantage of the polarized proton target is the possibility of producing polarized neutron beams in a wide energy range, from thermal to 10^5 eV. Using the time-of-flight technique with neutrons at the Pulsed Fast Neutrons Reactor (IBR), it was possible to rather easily study the spin states of individual resonances in neutron cross sections.

In this work, protons were polarized by the solid effect [92] in a $(La,Nd)_2 Mg_3(NO_3)_{12} \times 24H_2O$ single crystal containing a 0.5–1 % admixture of ^{142}Nd nuclei, at a temperature of 1–1.5 K in a 8 kA/cm magnetic field. Polarized neutrons obtained in this system passed through a polarized target made of polycrystalline metallic holmium. The ^{165}Ho nuclei were polarized by the Gorter-Rose method at 0.35 K in a 12 kA/cm magnetic field. This work yielded the spins of 23 levels of the ^{166}Ho compound nucleus [10].

The experiment which eliminated the uncertainty in the scattering lengths of neutrons on deuteron was carried out in the same laboratory [93]. If neutrons are scattered by nuclei far from resonances, the dependence of the interaction on spins manifests itself in the difference between the scattering lengths a_+ and a_- [34]. Two allowed sets of scattering lengths are found by measuring the coherent nuclear scattering length a and the total scattering cross section σ_S of nonpolarized neutrons on deuterons [29]. By virtue of the relation $\sigma_\pm = 4\pi a_\pm^2$, the formula (3.6) shows that the experimental identification of which set is actually realized requires that we determine the sign of σ_p in the energy range in which the coherence of neutron scattering by the target is negligible.

The authors of [93] measured the transmission of polarized neutrons by a polarized deuteron target. Deuterons were polarized by the same technique that was used for polarizing the protons in the deuterated lanthanum-magnesium nitrate curystal. Two runs of measuring the energy dependence of the relative change in transmission were carried out. The subsequent analysis of the results gave a set of scattering lenghts in which $a_+ > a_-$.

At the end of the 1970s, experiments were conducted at the NPL of JINR on the transmission of polarized neutrons by polarized targets in the energy range up to 10^5 eV [54, 94] (Fig. 3.1). This work is reviewed in [95]. The measurements were conducted by the time-of-flight technique at a pulsed IBR-30 reactor. The characteristics of the apparatus are given in Table 1.1 (see entry 10). As in the earlier work, the polarizer was a lanthanum-magnesium nitrate crystal. The polarization of neutrons could be reversed by rotating the polarized proton target together with the cryostat and the magnet 180° around the vertical axis. Nuclear targets were polarized by placing the specimens in a 12 kA/cm magnetic field and cooling them to ultralow temperatures, 0.03–0.04 K, by the Neganov method based on dissolving ^3He in ^4He [96].

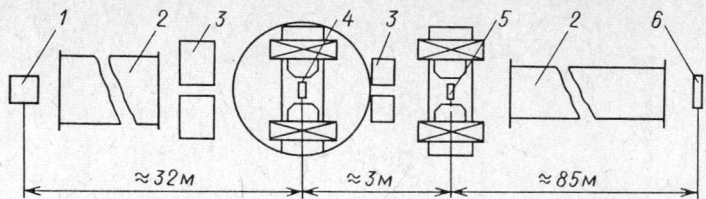

Fig. 3.1. Experiment on polarized neutron transmission through polarized targets [95]: *(1)* reactor, *(2)* evacuated tubes, *(3)* collimators, *(4)* polarized proton target, *(5)* polarized nuclear target, *(6)* neutron detector

Metallic terbium, holmium, and erbium, characterized by high internal magnetic fields of the order of 10^6–10^7 A/cm, and intermetallic compounds $TmFe_2$ and $PrAl_2$, which are ferromagnetic at ultralow temperatures, were used as specimens. Neutrons transmitted by the targets were recorded by a scintillation detector placed at a distance of 116 m from the reactor. Figure 3.2 shows the relative change in terbium transmission in the energy range of resolved resonances. The sign of relative change in transmission directly defines the value of spin J_c of the compound nucleus. All in all, the spins of about 250 resonances were identified in the five nuclei investigated, mostly for the first time.

The measurements of the relative transmission ε in the energy range of averaged cross sections in which resonances are not resolved yield even richer information. This information describes the energy range in which the number

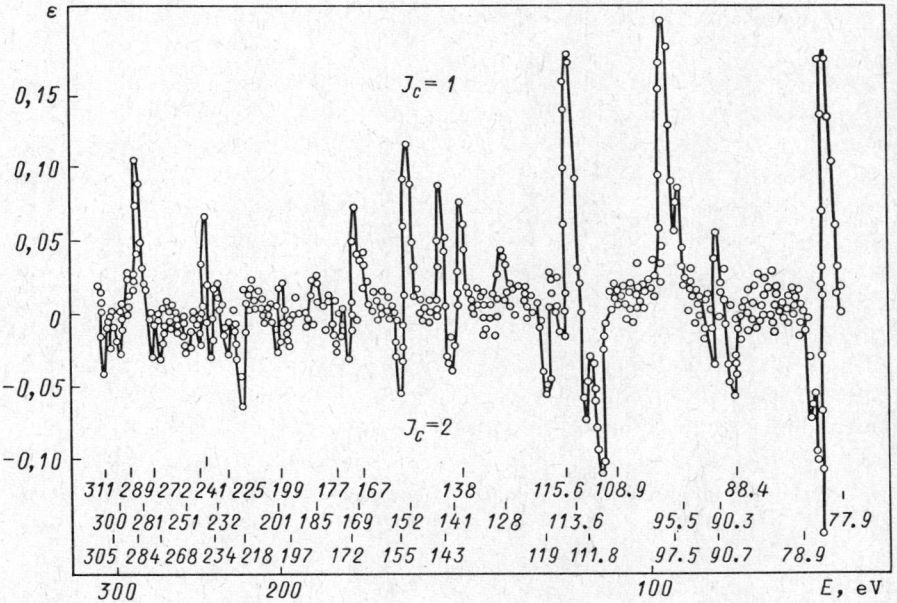

Fig. 3.2 Relative changes in transmission ε in a terbium specimen. The numerals below the curve give the resonance energy in electron volts

of resonances is so high that the error related to fluctuations is substantially reduced. In this way, data were obtained on the energy dependence of the difference between the strength functions of two spin states, $S_0^+ - S_0^-$, of the five nuclei investigated. Within the measurement error, this difference was found to be zero for all nuclei except holmium.

Study of Fissionable Nuclei. Experiments analyzing the parameters of resonances of fissionable nuclei, carried out at Oak Ridge National Laboratory with the neutron beam of the ORELA pulsed electron linear accelerator [97], deserve a separate description. The neutrons were polarized by the same method as at the NPL of JINR, i.e., by passing them through a polarized proton target. The investigated nuclei ^{235}U and ^{237}Np were polarized by magnetic hyperfine splitting in the ferromagnetic compounds US and NpAl$_2$, cooled to 0.6 K by the Neganov technique. Neutron transmission and the number of fission events in the specimen were found by measuring the fission neutron yield. Mean parameters of the resonances of ^{235}U and ^{237}Np were determined for the system of resonances with $J_c = J_i \pm 1/2$, in the energy range up to 25 keV.

The polarization of neutrons at Brookhaven National Laboratory was implemented by the diffraction method [98]. The ^{235}U nuclei were polarized by the same technique as in the earlier work [97], with US being cooled down to 0.1 K. The effect of transmission was measured, yielding the spins of 15 resonances of ^{235}U nuclei in the energy range 0.1–15 eV.

3.3 Nuclear Precession of Neutrons

The precession of the neutron spin in a beam of polarized slow neutrons passing through a polarized nuclear target (nuclear precession of neutrons) was predicted by *Baryshevsky* and *Podgoretsky* in 1964 [99]. This phenomenon consists in a rotation of the polarization direction of a polarized neutron beam around the direction of polarization in the nuclear target. The rotation is caused by the difference in nuclear scattering lengths for different spin states of the neutron–nucleus system. The rotation angle $\Delta\varphi$ is determined by the lengths of scattering of neutrons, $b_\pm$, on bound target nuclei in the states with spins $J_c = J_i \pm 1/2$:

$$\Delta\varphi = \frac{4\pi N d P_N}{k} \frac{J_c}{2J_c + 1}(b_+ - b_-) , \qquad (3.9)$$

where N is the number of nuclei per 1 cm^3, d is the length of the nuclear target, and $k = 2\pi/\lambda = mv/\hbar$ is the neutron wave number. The angular frequency of rotation of the polarization direction is given by the expression

$$\omega_0 = \frac{\Delta\varphi}{\Delta t} = \Delta\varphi \frac{v}{d} = \frac{4\pi N \hbar P_N}{m} \frac{J_c}{2J_c + 1}(b_+ - b_-) . \qquad (3.10)$$

The rotation frequency is thus independent of the velocity v (for slow neutrons); i.e., this frequency is a constant, characterizing a given material.

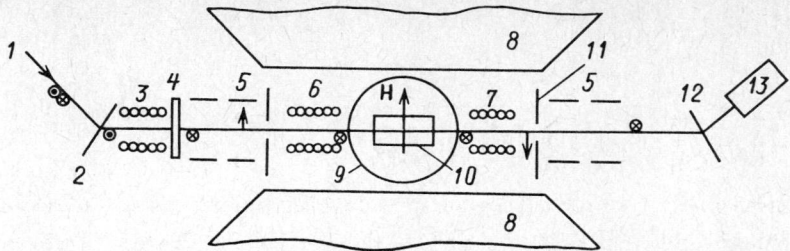

Fig. 3.3 Experiment measuring the spin-dependent scattering lengths by nuclear precession of neutrons [100]: *(1)* neutron beam, *(2)* spin polarizer, *(3)* spin flipper coil, *(4)* monitor, *(5)* leading magnetic field, *(6.7)* rf coils, *(8)* magnet, *(9)* Dewar flask, *(10)* target, *(11)* cadmium slit, *(12)* spin analyzer, *(13)* neutron counter; dots, crosses in circles, and arrows indicate the direction of neutron spins

The analyzer of neutron beam polarization can be used for measuring the rotation of the polarization direction. As a result, it is possible to find experimentally the difference $(b_+ - b_-)$ and (using additional information) the values of b_+ and b_-. The experiments for measuring b_+ and b_- by this method have been carried out since 1972 at the Centre d'Etudes Nucléaires at Saclay (France) [100] (Fig. 3.3). A neutron beam from the EL3 reactor was polarized by a CoFe single crystal which transmitted neutrons with a wavelength of 0.1074 nm and a polarization of P_{n1}. The beam then passed through a spin flipper, a leading magnetic field, and a polarized target placed in the magnet gap $(H = 20\,\text{kA/cm})$. Two rf coils were installed along the beam path, each rotating the neutron spins by 90°. Coil 6 rotated the spins perpendicularly to the magnetic field on the target (the vector $\boldsymbol{H}$ in Fig. 3.3). Hence, the spins precessed around this direction on the path from coil 6 to coil 7. The distance between the coils was adjusted so as to make an integral number of spin revolutions in the absence of a target and to have, at the entrance into coil 7, the same orientation that the spins had when leaving coil 6. Having traversed coil 7, the neutron spins were antiparallel to the field of the magnet. Leaving the second leading field, the spins were additionally rotated 90° to the direction coinciding with the orientation of spins after passing through the spin flipper. The analyzer, i.e., the second CoFe crystal, made it possible to measure the neutron beam polarization in precisely this direction, with the spin flipper first switched on and then off. The polarization was thus determined by the double reflection method [4].

An additional rotation of polarization direction in the target through an angle $\Delta\varphi$ reduced the measured polarization to a value $P_{n1}\cos\Delta\varphi$. Using the formulas of the double reflection method for calculating polarization, it was found (see Sect. 1.1.3) that

$$\cos \Delta\varphi = (R' - 1)/[P_{n1}P_{n2}(R' + 1)] \;, \tag{3.11}$$

where R' is the polarization ratio, and P_{n2} is the polarization efficiency of the analyzer. Equation (3.9) then yielded the sought difference between scattering lengths, $b_+ - b_-$. The differences $b_+ - b_-$ were found in this way for 27 nuclei [100].

4. Anisotropy of Gamma Rays Emitted by Polarized Nuclei After Polarized Neutron Capture

Experiments on the capture of polarized neutrons by polarized nuclei form a special class of experiments which make it possible to measure the spins of nuclear states. Using polarized neutrons and polarized nuclei, one can determine the relative contributions of two allowed states of the levels of compound nuclei with the spins $J_c = J_i \pm 1/2$ to the individual γ-transitions which follow immediately after the neutron capture (the so-called primary γ-transitions). This determination is similar to measuring the contributions of the states with the spins $J_i \pm 1/2$ to the total cross section in the experiments on the transmission of polarized neutrons through polarized nuclei (see Chap. 3).

The angular distribution of γ-quanta emitted after the capture of polarized neutrons by polarized nuclei also permits the determination of the spins of the final states of the levels of compound nuclei, J_f, provided the electromagnetic nature of the γ-transitions is an established fact.

4.1 Fundamentals of the Experimental Method

Let us assume that the capture of polarized neutrons by nuclei with spins J_i occurs in the s state. The spins of the compound nuclei are then allowed to have two values: $J_c = J_i - 1/2$ and $J_c' = J_i + 1/2$. These states (spin channels) can also occur simultaneously, as a result of overlapping of the resonances which determine the contribution to the thermal capture cross section. Compound nuclei chiefly emit primary high-energy γ-quanta, increasing the population of the lower energy levels of nuclei. We shall restrict the analysis to the multipoles L and L' of the primary γ-radiation which populates the level with the spin J_f. The scheme of the relevant (n, γ) reaction is shown in Fig. 4.1.

The general expression giving both the angular distribution of the emitted gamma quanta and their circular polarization (this aspect will be discussed in Chap. 5) is [101]:

$$I(\theta) = J\sigma_a N d(\Gamma_f/\Gamma_{\text{tot}}) S_0 \{1 + \sum_{kk_1k_2} \varepsilon_k A_k^{k_1 k_2} f_{k_1}(n)$$
$$\times f_{k_2}(J_i) P_k(\cos\theta)\} , \qquad (4.1)$$

where J is the neutron flux in neutrons/s, σ_a is the capture cross section of

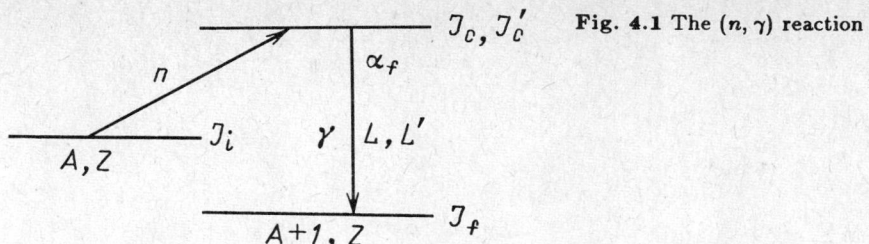

Fig. 4.1 The (n, γ) reaction

nonpolarized neutrons by a nonpolarized target, Nd is the thickness of the nuclear target, cm^{-2}, Γ_f is the partial radiation width of the primary γ-transition to the level with the spin J_f, Γ_{tot} is the total radiative width, S_0 is the efficiency of the detector which measures the γ-emission at an angle θ to the direction of nuclear polarization, ε_k is the parameter of efficiency of the apparatus, $f_{k_1}(n)$ and $f_{k_2}(J_i)$ are the orientation parameters of the neutron beam and the nuclear target (see Sect. 3.1; it is assumed here that the beam and the target have a common orientation axis); $P_k(\cos\theta)$ are the Legendre polynomials, and $A_k^{k_1 k_2}$ are numerical coefficients. The summation in (4.1) is subjected to the following constraints on the integral values of k, k_1, and k_2:

$$0 \leq k_1 \leq 1 \ ; \quad 0 \leq k_2 \leq 2J_i \ ; \quad |k_1 - k_2| \leq k \leq k_1 + k_2 \ ; \quad k + k_1 + k_2 \text{ is even}.$$

The efficiency parameter ε_k of the apparatus is equal to unity for even k and to the efficiency of detecting the circularly polarized γ-emission, ε_c, for odd k. The formula (4.1) gives the efficiency of detecting the γ-emission from a thin target ($\sigma_a Nd \ll 1$). The general expression for the coefficients $A_k^{k_1 k_2}$ is

$$A_k^{k_1 k_2} = \sum_{J_c J_c' L L'} (-1)^{J_f - J_c'} J_i^{k_2} \binom{2k_2}{k_2}$$

$$\times \sqrt{\frac{(2k_2 + 1)(2J_i - k_2)!}{2(2J_i + k_2 + 1)!}} \hat{J}_c \hat{J}_c' \hat{k}_1 \hat{k}_2 C(k_2 0 k_1 0 | k0)$$

$$\times \begin{Bmatrix} J_i & \frac{1}{2} & J_c \\ J_i & \frac{1}{2} & J_c' \\ k_2 & k_1 & k \end{Bmatrix} \bar{z}_1(LJ_c L' J_c'; J_f k) \eta^r / \left(\frac{J_i}{2J_i + 1} + \frac{J_i + 1}{2J_i + 1} \eta^2 \right),$$

(4.2)

where $\hat{k} = \sqrt{2k+1}$; $C(k_2 0 k_1 0 | k0)$ are the Clebsch-Gordan coefficients, $\begin{Bmatrix} \cdots \\ \cdots \\ \cdots \end{Bmatrix}$ are the Wigner symbols, and

$$\bar{z}_1(LJ_c L' J_c'; J_f k) = (-1)^{k-L+L'-1} \hat{L}\hat{L}' \hat{J}_c \hat{J}_c' C(L1L' - 1|k0)$$
$$\times W(LJ_c L' J_c'; J_f k) \ ;$$

$W(LJ_c L' J_c'; J_f k)$ are the Racah coefficients (see [102]).

In (4.2), $r = 0, 1, 1$, and 2 for the combinations $J_c J_c$, $J_c J_c'$, $J_c' J_c$, and $J_c' J_c'$, respectively. The parameter $\eta = a_+/a_-$ is the ratio of channel amplitudes of neutron capture into the spin states $J_i + 1/2$ and $J_i - 1/2$.

Let us introduce the parameter of mixing of spin channels,

$$\alpha_f = \sigma_f^+/(\sigma_f^+ + \sigma_f^-) ,$$

where $\sigma_f^\pm$ is the partial cross section determining the contribution of the channel $J_i \pm 1/2$ to the total cross section σ_f, which describes the population of the level with the spin J_f. Taking into account that [34]:

$$\sigma_f^+ = \frac{J_i + 1}{2J_i + 1} a_+^2 ; \quad \sigma_f^- = \frac{J_i}{2J_i + 1} a_-^2 ,$$

we find

$$\alpha_f = (J_i + 1)\eta^2/[J_i + (J_i + 1)\eta^2] . \tag{4.3}$$

The mixing parameter α_f varies from 0 to 1 because η can assume any real value.

The parameters $A_k^{k_1 k_2}$ can be rewritten in the form

$$A_k^{k_1 k_2} = a\alpha_f + b(1 - \alpha_f) \pm c\sqrt{\alpha_f(1 - \alpha_f)} . \tag{4.4}$$

The first term in (4.4) represents the spin channel $J_i + 1/2$, the second term represents the channel $J_i - 1/2$, and the third represents the interference of these channels. The plus sign in front of the third term signifies, by convention, the constructive interference, while the minus sign signifies the destructive interference. The evaluation of the coefficients $A_k^{k_1 k_2}$ from the experimental data makes it possible to find the parameters of the levels of compound nuclei.

In the analysis below we use a simplifying assumption of the purely dipole nature of the primary γ-emission, and consider all orientation parameters, except the first, to be negligible. This means that $f_1(n) = P_n$, and $f_1(J_i) = P_N$. Equation (4.1) is thereby simplified:

$$I(\theta) = I\sigma_a N d(\Gamma_f/\Gamma_{tot}) S_0 \{1 + P_n P_N [A_0^{11} + A_2^{11} P_2(\cos \theta)]\} . \tag{4.5}$$

The coefficients A_0^{11} and A_2^{11} are conveniently found if the intensities $I(\theta)$ are measured in two cases: (1) the directions of polarization of the neutron beam and the target are parallel, $I_{\uparrow\uparrow}(\theta)$, and (2) these directions are antiparallel, $I_{\uparrow\downarrow}(\theta)$. The relative intensity difference is usually denoted by $e(\theta)$:

$$e(\theta) = [I_{\uparrow\uparrow}(\theta) - I_{\uparrow\downarrow}(\theta)]/[I_{\uparrow\uparrow}(\theta) + I_{\uparrow\downarrow}(\theta)] . \tag{4.6}$$

In order to increase the intensity of γ radiation, thick targets are often used in experiments with polarized nuclei. Consequently, neutron polarization is not constant along the beam in the target, owing to scattering and depolarization processes; this fact can be accounted for by introducing the effective neutron polarization P_n^*. We find

$$e(\theta) = P_n^* P_N \{F(d)\varrho + A_0^{11} + A_2^{11} P_2(\cos \theta)\} , \tag{4.7}$$

where $F(d)$ is a factor which takes into account the target thickness, $\varrho = \sigma_p/\sigma_a$ is the spin-dependent factor in the total capture cross section, and σ_p is the polarization cross section (see Sect. 3.1). The factor $F(d)$ tends to 1 for an infinitely thick target, and to 0 for an infinitely thin one.

The γ emission intensity is measured for $\theta = 0$ and $\theta = 90°$, yielding the coefficients A_0^{11} and A_2^{11}:

$$A_0^{11} = F(d)\varrho - [e(0) + 2e(90°)]/3P_n^* P_N \; ;$$

$$A_2^{11} = 2[e(0) - e(90°)]/3P_n^* P_N \; .$$
(4.8)

4.2 Survey of Experiments

The first measurement of the angular distribution of γ quanta emitted by polarized nuclei after a capture of polarized neutrons were carried out in 1973 by two groups: Postma's group at the Netherlands Energy Center (ECN) at Petten [103] and Honzatko's group in Řež (Czechoslovakia) [104]. The former worked with polarized ^{143}Nd, ^{145}Nd, ^{147}Sm, ^{149}Sm, and ^{152}Sm nuclei, the latter with ^{59}Co nuclei.

In 1977 the system at Petten has been modified [105] (Fig. 4.2). Identical CoFe single crystals (92 % Co, 8 % Fe), mounted in permanent magnet

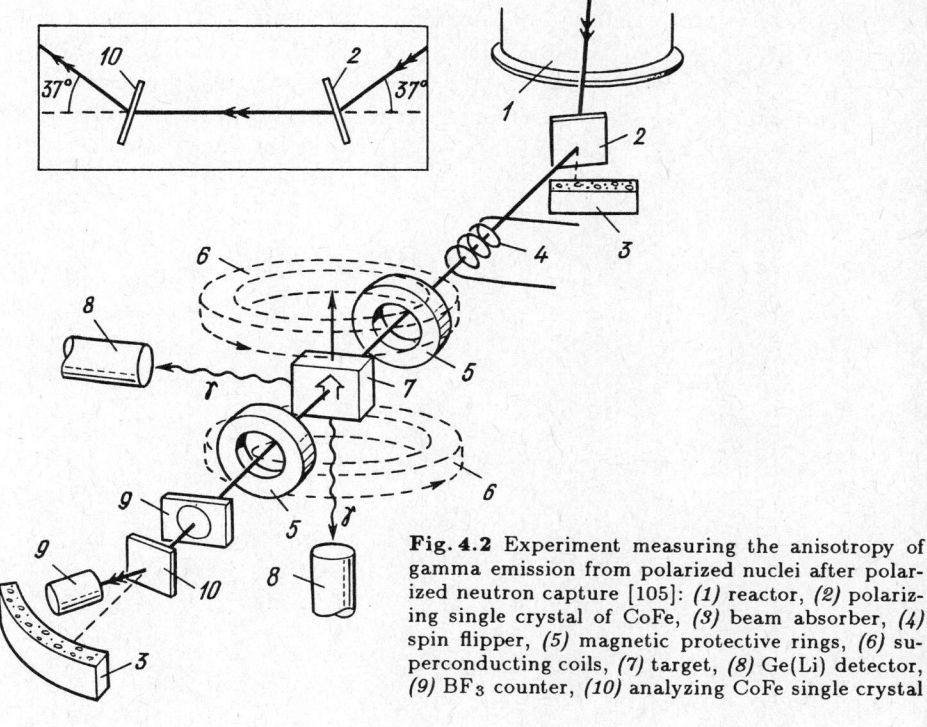

Fig. 4.2 Experiment measuring the anisotropy of gamma emission from polarized nuclei after polarized neutron capture [105]: *(1)* reactor, *(2)* polarizing single crystal of CoFe, *(3)* beam absorber, *(4)* spin flipper, *(5)* magnetic protective rings, *(6)* superconducting coils, *(7)* target, *(8)* Ge(Li) detector, *(9)* BF$_3$ counter, *(10)* analyzing CoFe single crystal

gaps, were used as the neutron polarizer and analyzer. The polarized beam of neutrons with an energy of 0.065 eV was formed at a scattering angle of 37° and (200) reflection in the single crystal. The neutron flux density and beam polarization at the target site reached 10^5 neutron/(cm^{2}s) and (97±1) %, respectively. Neutron spins were reversed every 100 s, at a probability close to 1, by a resonance rf spin flipper (see Sect. 1.2.3). Neutrons were counted by a BF$_3$ detector, and γ-quanta were recorded by means of two Ge(Li) detectors with a resolution of 2.2 and 5.6 keV at the energies of 1.33 and 7 MeV, respectively. The axes of the two detectors were perpendicular to the neutron beam (Fig. 4.2). The background of γ-ray emission due to neutron scattering in the target was reduced by surrounding the target with a 5–6 mm layer of ^{6}LiF, and making the cryostat walls, closest to the target, of aluminum. A lead shield, 10 cm thick, protected the system from external gamma radiation.

The cryogenic equipment for the polarization of target nuclei consisted of two main parts installed one above the other. The upper part was a refrigerator with a solution of ^{3}He in ^{4}He, and the lower part was a cryostat with two superconducting coils. The target placed between these coils had a maximum size of 2.5×2.5×1 cm. The maximum field on the target was 40 kA/cm, and the target temperature under operating conditions was (30–50)×10^{-3} K.

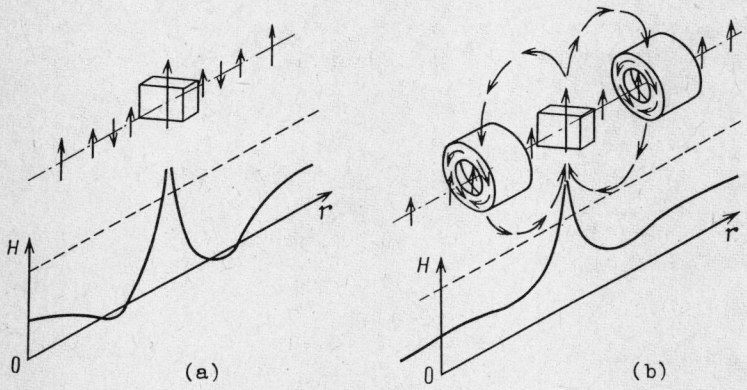

Fig. 4.3a,b. Magnetic field along the beam in the region of cryogenic equipment [105]: (a) without magnetic protective rings, (b) with magnetic protective rings; in case (a), there are regions with reversed direction of the magnetic field

Unless special protective measures were taken, neutron depolarization could be produced in the region of reverse field of superconducting coils, since the region of zero field was present (Fig. 4.3). This effect was eliminated by surrounding the beam with two protective rings made of Armco steel. By using permanent magnets and iron strips, it was possible to compensate for the reverse field and achieve a field of constant sign along the entire path of neutrons.

The above-described apparatus was used to study the ^{59}Co$(n,\gamma)^{60}$Co reaction. Altogether, 36 γ-transitions were studied. Admissible intervals of α_f and the spins J_f of the levels of ^{60}Co were obtained for these transitions.

In 1979, the CoFe single crystals were replaced by Heusler alloy (Cu_2MnAl) single crystals providing a greater flux density of polarized neutrons (see [29]). The parameters of ^{27}Al, ^{55}Mn and ^{141}Pr nuclei were then measured [106].

4.3 Coherent Interference of Spin States

As follows from (4.1) and (4.2), the angular distribution of γ-radiation contains a term corresponding to the coherent interference of spin states

$$J'_c = J_i + 1/2 \quad \text{and} \quad J_c = J_i - 1/2 \ .$$

This term can be measured if either neutrons or nuclei are polarized ($k_1 \neq 0$ or $k_2 \neq 0$).

Honzatko et al. [104] were the first to single out this effect in the measurements of the angular distribution of γ quanta emitted in the capture of polarized neutrons by polarized nuclei of ^{60}Co. The coherent interference of two allowed spin states in neutron capture plays an important role in the interpretation of experimental results. In the analysis of experimental data, the obtained relative differences between $e(0)$ and $e(90°)$ for different energies of γ-quanta were compared with the theoretical values of these quantities. These values are functions of the ratio $\eta^2/(1+\eta^2)$ and of the spins of final states, J_f. The experiment identified the parameters η and the allowed spins J_f for 20 levels of ^{60}Co.

In [105], the number of investigated levels of ^{60}Co was increased to 36, and some ambiguity in spin values was eliminated.

5. Circular Polarization of Gamma Rays Emitted by Nuclei After Polarized Neutrons Capture

Another phenomenon which makes it possible to measure the spin states of the levels of compound nuclei is the circular polarization of γ-quanta emitted after a nucleus captures a polarized neutron. Measurements of circular polarization also yield, in some cases, the characteristics of the gamma radiation emitted by nuclei.

5.1 Fundamentals of the Experimental Method

The projection of the angular momentum of a γ quantum on the direction of its momentum (the so-called helicity λ) can assume only two values: $\lambda = \pm \hbar$. A quantum with $\lambda = +\hbar$ is said to have right-handed circular polarization, and a quantum with $\lambda = -\hbar$ is said to have left-handed circular polarization. The right(left)-handed circular polarization of a γ-quantum corresponds to the notion of clockwise (counterclockwise) rotation of the vectors of electric and magnetic field strength in the classical electromagnetic wave for the observer who looks along the direction of wave propagation.

By definition, the circular polarization of a beam of γ-quanta is the ratio

$$P_\gamma = (N_R - N_L)/(N_R + N_L) , \tag{5.1}$$

where $N_{R(L)}$ is the number of right(left)-handed circularly polarized γ-quanta in the beam.

The efficiency of a detector with respect to circularly polarized γ-quanta is described by the terms with odd k in (4.1). If polarized neutroons are captured by nonpolarized nuclei, the expression (4.1) becomes simpler:

$$I(\theta) = \text{const}\{1 + \varepsilon_c f_1(n) A_1^{10} \cos \theta\} . \tag{5.2}$$

where ε_c is the efficiency of detecting circularly polarized gamma quanta, $f_1(n) = P_n$ is the neutron beam polarization, and θ is the angle between the momentum of a gamma quantum and the vector of polarization of the neutron. The quantity $R = A_1^{10}$, called the polarization function, depends both on the mixing parameter of spin channels, α_f, given by (4.3), and on the mixing parameter δ of the multipoles of gamma radiation. A complete analysis being beyond reach, one either introduces simplifying assumptions on the purely dipole nature of emission or on a single spin channel, or one conducts additional experiments for measuring the characteristics of the same γ-transitions by different methods.

Several methods available for measuring the circular polarization of γ-quanta are reviewed in [107]. The choice of method is very dependent on the energy of the studied γ-quanta. In the high-energy range (3–10 MeV) typical for neutron capture, the most suitable technique is the transmission of γ-quanta through a magnetized ferromagnet (polarimeter). This method is based on the strong dependence of the Compton scattering cross section on the angle between the vectors of the γ-quantum momentum and the electron spin. If the angle between them is 0 or 180°, the total cross section of the Compton scattering of circularly polarized γ-quanta in a magnetized ferromagent can be written in the form

$$\sigma = \sigma_0 \pm f P_\gamma \sigma_c , \tag{5.3}$$

where σ_0 is the cross section independent of the polarization of γ-quanta; f is the fraction of polarized electrons in the ferromagnet; the plus (minus) sign corresponds to coinciding (opposite) signs of the projection of angular momentum (helicity) and projection of the electron spin on the direction of momentum of the γ quantum;

$$\sigma_c = 2\pi r_0^2 \left[\frac{1 + 4k_0 + 5k_0^2}{k_0(1+2k_o)^2} - \frac{1+k_0}{2k_0^2} \ln(1+2k_0) \right] ; \tag{5.4}$$

r_0 is the "classical radius" of the electron; and $k_0 = E_\gamma/m_0 c^2$ is the energy of the γ quantum in units of the rest energy of the electron.

The relative difference between detector counting rates is related to the circular polarization by the formula

$$P = \frac{I(0) - I(180°)}{I(0) + I(180°)} = \varepsilon_c P_n R \approx -nl\nu\sigma_c P_\gamma , \tag{5.5}$$

where n is the number of atoms of the ferromagnetic scatterer per unit volume, l is the scatterer length, and $\nu = fZ$ is the mean number of polarized electrons per atom (in permendur, $\nu \approx 2.5$).

The value of P is not high because the fraction f of polarized electrons does not exceed 10%. The length of the scatterer cannot be considerably increased because it results in a parallel increase in the absorption of γ-rays. The scatterer length is chosen so as to minimize $\Delta P/P$, where ΔP is the statistical error of measuring P. This condition prescribes the optimal length of the scatterer [107]:

$$l_0 = 2/n\tau , \tag{5.6}$$

where τ is the total coefficient of absorption of gamma radiation by one atom of the scatterer.

5.2 Survey of Experiments

The circular polarization of the gamma radiation emitted by ^{57}Fe and ^{64}Cu compound nuclei after polarized neutron capture was first measured by *Trumpy* in 1955 [108]. The polarized neutron beam was obtained by passing the beam through magnetized iron. The polarization of gamma quanta was measured by passing them through analyzers also made of magnetized iron. Later Trumpy studied the compound nuclei of ^{33}S, ^{41}Ca, ^{49}Ti, ^{54}Cr, ^{57}Fe, ^{59}Ni, ^{65}Zn, and ^{183}W [109]. The relative difference in counting rates, P, did not exceed 0.05 %.

In 1962, *Michalec* and *Ruskov* [110] somewhat modified Trumpy's apparatus. Polarized neutrons were obtained by reflection from a cobalt mirror. The magnets analyzing the circular polarization of gamma radiation were made of vanadium permendur. The background effect was reduced drastically. As a result, it was possible to increase the experimental effect by an order of magnitude in comparison with pioneer publications [108, 109]. The compound nuclei studied in [110] were ^{57}Fe, ^{59}Ni, and ^{64}Cu.

In 1961, *Vervier* [111] reported that he was able to measure the circular polarization simultaneously in several γ transitions. Later *Kopecky* et al. [112] published data on the compound ^{33}S, ^{41}Ca, ^{60}Co, ^{64}Cu, and ^{92}Zr nuclei.

Beginning in 1965, *Abrahams* and his coworkers [113, 114] started a thorough and systematic study of circular polarization of the γ-emission of a large number of nuclei at the high-flux reactor at Petten; the work was aimed at determining the spins of the levels and the miltipolarity of transitions. All in all, several hundred γ-transitions were analyzed. By using Ge(Li) detectors for detecting the gamma radiation, it was possible to obtain detailed information in a wide range of energies of γ-quanta.

We shall describe in detail the modified apparatus at Petten [114] (Fig. 5.1). The parameters of the polarized neutron beam formed by a multislit reflecting collimator comprising 74 permendur mirrors are given in Table 1.1 (entry 6). The circular polarization of γ-rays was measured by two polarimeters made of magnetized pemendur cylinders, 60 mm in diameter, 80 mm long, followed by Canberra Ge(Li) detectors with a registration efficiency of 15 % and a resolution of 7.5 keV for 7.0 MeV gamma quanta. The pulses of each Ge(Li) detector were fed to one of the four quadrants of the 4096-channel pulse analyzer.

Neutron spins were flipped every 100 s for averaging out neutron flux fluctuations and circuit drifts; the spins were at the angles $\theta = 0$ or $180°$ to the momenta of γ-quanta. The adiabatic spin rotation was effected by a system of two magnetic guides consisting of successively switched coils which rotated the spins by 1/4 of a right- or leftward revolution around the momentum direction. Correspondingly, the circular polarization of each γ-line reversed its sign every 100 s. The multichannel analyzer which accumulated the spectra from the two detectors switched at the same period either to subtracting the spectra, or to adding them. The difference spectra of the detectors were accumulated in two 1024-channel quadrants of the analyzer memory. The sum spectra were accumulated for 1/4 of the total exposure time in the other two quadrants.

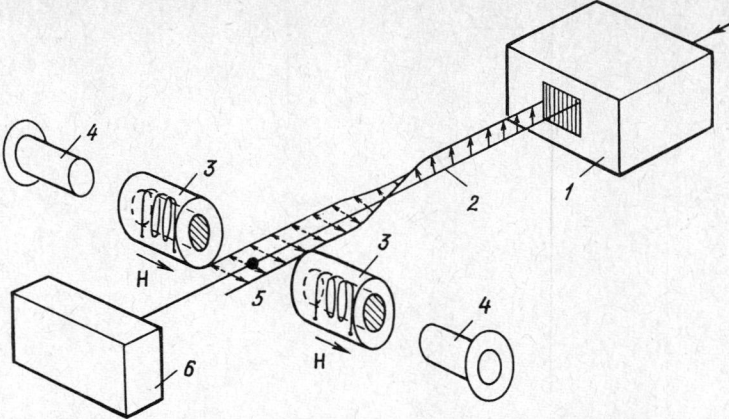

Fig. 5.1. Device for measuring the circular polarization of gamma rays emitted from nuclei after polarized neutron capture [114]: (1) magnetized permendur mirrors, (2) the system for rotating the spin either to the right (*dashed arrows*) or to the left (*solid arrows*), (3) polarimeters made of magnetized permendur cylinders, (4) Ge(Li) gamma-ray detectors, (5) target, (6) beam trap

Every 100 s, the analyzer data were written into the magnetic disk of the computer. The spectra were then summed up. Every 24 hours, the two spectra corresponding to the two spin orientations were corrected for the drift of electronic circuits, after which the final summation of spectra took place. Standard measurement runs with one target lasted about 300 hours.

In order to extract the polarization function from the final spectra, it is necessary to know how polarimeter efficiency depends on the γ-quantum energy (Fig. 5.2). The absolute calibration was made with 5.42 MeV γ-quanta emitted in the reaction $^{32}S(n,\gamma)^{33}S$ for which R is known: $R = -0.5$.

The targets must be free of hydrogen, otherwise noncoherent scattering and depolarization of polarized neutrons takes place. For this reason, mostly pure dry oxides of elements in thin aluminium or Teflon containers were used.

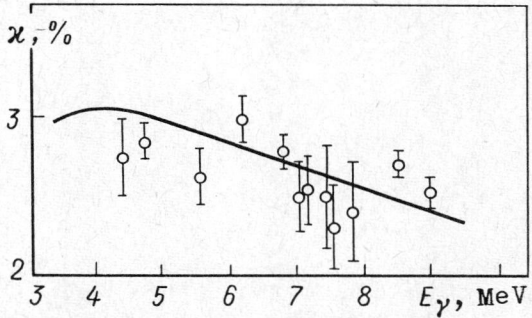

Fig. 5.2. Measured efficiency $\varkappa$ of polarimeters as a function of the energy of gamma quanta. Circles show experimental data; the solid curve represents theoretical prediction

An experiment at LINP [115] measured the circular polarization of gamma quanta produced in the radiative capture of thermal polarized neutrons by protons. The experiment was conducted to determine the possible admixture of the triplet capture state $^3S_1$ to the principle capture state $^1S_0$ in the reaction $np \rightarrow d\gamma$. Solid parahydrogen ($300\,\text{cm}^3$) was used as the target. The circular polarization was found to be $P_\gamma = -(2.90\pm0.87)\times10^{-3}$. Taking into account the contribution of the known admixture of the D wave to P_γ in the initial and final states of the n-p system, it is concluded that the relative contribution of the amplitude of transitions from the $^3S_1$ state to the radiative np-capture cross section is very small: $\sigma(^3S_1)/(\sigma(^1S_0)\leq 8\times10^{-6}$.

6. Quantum-Mechanical Symmetry Properties and Polarized Neutrons

Ever since *Wu* et al. [116] discovered that the weak interaction violates parity symmetry, it has been clear that polarized neutron beams constitute a potentially powerful tool in the search for violations of the quantum-mechanical conservation laws.

The experiments measuring correlations in the beta decay of free polarized neutrons created the foundation for the entire edifice of weak interaction theory. The neutron participates simulatenously in both the weak, electromagnetic, and strong interactions, thus opening new possibilities for the study of nonconservation of P parity in nuclear interactions and for the search for violations is of time reversal symmetry in nuclear interactions.

The chapters that follow give a detailed analysis of the application of polarized neutrons to the solution of these problems. The present chapter recapitulates for the reader the essential facts of quantum-mechanical conservation laws and their violation in the weak interaction; it also outlines some aspects of weak interaction theory which are essential for the understanding of experiments with polarized neutrons.

6.1 Quantum-Mechanical Conservation Laws

Quantum mechanics deals with three discrete transformations and three conservation laws stemming from these transformations [1]:

1. inversion of coordinates, i.e., simultaneous reversal of signs of three spatial coordinates: $x \to -x$, $y \to -y$, $z \to -z$; this transformation is denoted by P;
2. time reversal, i.e., the reversal of sign of time: $t \to -t$; it is denoted by T; and
3. charge conjugation, i.e., replacement of all particles by their antiparticles, e.g., the reversal of the sign of charge of all charged particles; it is denoted by C.

We recall how certain physical quantities, such as coordinate vector r, momentum vector p (true vectors), spin vector s (or J) (axial vector), and their products (scalar and vector products), are transformed under the P and T transformations (Table 6.1) [34].

Table 6.1. Transformations of some quantities under coordinate inversion and time reversal

Transformation	r	p	p^2	s	s^2	sp	$s(p_1 \times p_2)$	$s(p \times J)$	$(p_1\, p_2)$	$J(p_1 \times p_2)$
P	$-r$	$-p$	p^2	s	s^2	$-sp$	$s(p_1 \times p_2)$	$-s(p \times J)$	$(p_1 p_2)$	$J(p_1 \times p_2)$
T	r	$-p$	p^2	$-s$	s^2	sp	$-s(p_1 \times p_2)$	$-s(p \times J)$	$-(p_1 p_2)$	$J(p_1 \times p_2)$

Note that pseudoscalar quantities such as sp reverse their sign under coordinate inversion, while a triple product such as $s(p_1 \times p_2)$ reverses its sign under time reversal but preserves it under coordinate inversion.

Each quantum-mechanical transformation implies a quantum-mechanical quantity which either reverses or preserves its sign under this transformation. The symmetry under discrete transformations implies certain quantum-mechanical conservation laws [1].

Inversion of coordinates corresponds to spatial parity (P parity) and the law of P parity conservation, time reversal corresponds to the invariance under time reversal (T invariance) and the law of T invariance, and charge conjugation corresponds to charge conjugation parity (C parity) and the law of C parity conservation.

By definition, the P parity of a quantum system can take on two values: $P = +1$, if the wave function of the quantum system preserves its sign under coordinate inversion, and $P = -1$, if its sign is reversed. The law of P parity conservation states that a quantum system can be only in a state with definite P parity: either $P = +1$ (even state) or $P = -1$ (odd state). If the system remains isolated, its P parity does not change in time.

The charge conjugation parity of a system ($C = \pm 1$) and the law of charge parity conservation (under the C transformation) are defined by analogy.

Relativistic quantum theory imposes the constraint that system be invariant under the transformation in which all three operations are carried out simultaneously: coordinate inversion (P), time reversal (T), and charge conjugation (C). This statement is called the Lüders-Pauli, or CPT, theorem [117]. The CPT theorem relates the time reversal transformation T to the CP transformation, called combined inversion and standing for the simultaneous coordinate inversion P and charge conjugation C. The CP transformation corresponds to the conservation of CP parity (combined parity).

It can be shown (see, e.g., [34]) that the symmetry under coordinate inversion is equivalent to the symmetry with respect to mirror reflection which reverses the sign of only one coordinate. Consequently, in the language of experimental physics, the law of P parity conservation signifies that it is always possible to produce a mirror copy of any physical instrument. A copy is bound to record the same quantilook at the copy in a mirror. In other words, P parity conservation states that if a certain process occurs in nature, then the mirror-reflected process occurs with the same probability. This formulation is known as the mirror symmetry of physical processes.

In mathematical terms, the P parity conservation law signifies that the physical quantities measured in an experiment cannot be combined to form a pseudoscalar quantity. As can be seen in Table 6.1, *sp*-type pseudoscalar quantities are sign-reversing under coordinate inversion.

6.2 Violation of P, C, and CP Parities in the Weak Interaction

For a long time, P parity conservation was regarded as a universal law, that is, as holding for any interactions of any physical objects. The universality of the mirror symmetry principle in microscopic phenomena ended in 1956, when *Lee* and *Yang* [118], analyzing the decay characteristics of K^+ mesons (the so-called θ-τ puzzle), demonstrated that different decays of K^+ mesons manifest properties opposite to those of mirror symmetry, and hence, that K^+ mesons do not possess a definite spatial parity. The decays of K^+ mesons are manifestations of the weak interaction (which is one of the four fundamental interactions). Lee and Yang conjectured on the basis of data analysis that P parity, as in the case of K^+ decays, is not conserved in all weak interaction processes.

Lee and Yang suggested a number of experiments to test their hypothesis. One of them was carried out by *Wu* et al. [116] at Columbia University. It was shown that spatial parity is not conserved in the β-decay of polarized ^{60}Co nuclei. Wu and her coworkers observed an asymmetry w in the direction of electron emission relative to the spin direction of the decaying polarized ^{60}Co nuclei. This asymmetry stems from a pseudoscalar quantity $\boldsymbol{Jp}$, where $\boldsymbol{J}$ is the spin of the ^{60}Co nucleus and $\boldsymbol{p}$ is the decay electron momentum; it is given by the formula

$$w = \text{const}(1 + a\boldsymbol{Jp}) \ . \tag{6.1}$$

The factor a is called the asymmetry coefficient.

Other experiments carried out shortly after that of Wu et al. conclusively demonstrated that the Lee-Yang hypothesis on the violation of P parity in the weak interaction is correct, and that C parity is also violated, simultaneously with P parity. *Landau* [64] then hypothesized that charge-mirror symmetry exists, i.e., that the weak interaction conserves CP parity.

However, in 1964, experiments on $K_2^0 \to 2\pi$ decays detected the violation of CP parity [65]. Several hypotheses pointed to either the intermediate strong [119] or the electromagnetic interaction [120] as the factor responsible for the nonconservation of CP parity. All these hypotheses predicted that CP parity, and by virtue of the CPT theorem, T invariance, can be violated in nuclear interactions.

It thus became necessary to test whether the quantum-mechanical conservation laws, connected with discrete transformations, are violated in other nuclear physics experiments as well. Polarized neutron techniques made such experiments possible.

6.3 The Role of Polarized Neutrons in the Investigation of Quantum-Mechanical Symmetry Properties

We have already mentioned that polarized neutron beams, using a direction specified in space by the axial neutron spin vector s, offer a unique opportunity of studying various correlations of physical quantities with respect to this direction.

This opportunity was first realized in experiments on the decay of polarized neutrons. The neutron is a convenient object for studying the correlations, because the theortical interpretation of decay is simpler for the neutron than for heavier nuclei, owing to the lack of nuclear structure. When free neutrons decay, it is possible to detect both decay electrons and recoil nuclei (i.e., protons), thus increasing the amount of extracted information.

The discovery of P parity nonconservation in the weak interaction stimulated a search for an interaction potential between nucleons which would not conserve P parity. Pauli was of the opinion that "the actual problem now seems to be the question why are strong interactions right and left symmetric" (see [121]). The fact that the neutron participates in both the electromagnetic and strong interactions makes it possible to carry out experiments searching for parity nonconservation in electromagnetic and strong interactions or, to be more exact, in nuclear interactions.

One of the corollaries of the hypothesis of the universality of the weak interaction, advanced in 1958 [122], was the hypothesis that the neutron and the proton, components of all nuclei, interact not only through strong and electromagnetic, but weak interactions as well. It was necessary to prove experimentally that nucleons are indeed coupled through the weak interaction; however, the weak coupling constant is much smaller than the strong coupling constant, so that the weak interaction effects had to be found against the much more intensive strong interaction background. Success was achieved only through utilizing a unique property of the weak interaction: nonconservation of P parity.

6.4 Some Aspects of the Weak Interaction Theory

Since the subsequent chapters are devoted to experiments with polarized neutrons, analyzing the properties of the weak interaction, this and the following subsections present a brief review of the fundamentals of weak interaction theory, sufficient for understanding the material that follows [58, 123].

The discussion will be restricted to processes occuring at low energies. In this case, the standard Weinberg-Salam electroweak model [82] describes weak processes by using an effective four-fermion Lagrangian containing two terms,

$$\mathcal{L}^w = \mathcal{L}^{ch} + \mathcal{L}^n,\tag{6.2}$$

where $\mathcal{L}^{ch}$ and $\mathcal{L}^n$ are the Lagrangians of interaction between charged and neutral currents. Both these Lagrangians consist of the products of two currents: the charged and the neutral, respectively [58].

Although in the general case the weak currents are coupled via the exchange of intermediate charged and neutral bosons, the low-energy weak processes can be considered to taking place without these intermediaries, so that one can speak of the local four-fermion interaction. For instance, the decay of the neutron can be represented by the diagram in Fig. 6.1.

Fig. 6.1. Decay of the neutron

Weak interaction theory has derived from the constraints of relativistic invariance that, in the general case, the charged current has a structure described by terms possessing specific symmetry properties; namely, the properties of a scalar (S), polar vector (V), tensor (T), axial vector (A), and pseudoscalar (P). The corresponding expressions are referred to as the variants of the weak interaction in general and of the β-decay of nuclei in particular. Each variant is characterized by the contribution of the corresponding term to the complete expression. The coefficients are denoted by C_S, C_V, C_T, C_A, and C_P, and are called the constants of the respective variants in the weak interaction. Each variant of nuclear β-decay has a corresponding nuclear matrix element determined by the properties of the wave functions of the initial and final states. In each variant, the maximum value of the matrix element, corresponding to an allowed transition, is characterized by specific selection rules for spin and parity that are imposed on the initial and final states of the nucleus.

According to the Fermi selection rules, a β-transition is allowed if it leaves both the spin and the parity of the nucleus unchanged ($\Delta J = 0$, $P_i/P_f = 1$). According to the Gamov-Teller selection rules, those β-transitions are allowed which leave the parity of the nucleus unchanged ($P_i/P_f = 1$), while the spin is allowed to change by $\Delta J = 0$, ± 1, except for the transition $J = 0 \to J = 0$. The corresponding matrix elements are denoted by M_F and M_{GT}.

More than 20 years of experimental work went into establishing the characteristics of the β-decay interaction, and a number of erroneous conclusions were drawn on the way. The notion that the S and T variants are realized in the β-decay was predominant for a long time, but soon after the discovery of parity violation in the weak interaction, new experiments were carried out, some of them involving the decay of polarized neutrons, which contradicted the S and T variants.

If P parity and T invariance are violated, in the general case the number of constants of the variants of the weak interaction is 20. Each constant C_j of a weak interaction variant corresponds to a constant C'_j appearing in the interaction Hamiltonian because of P parity violation. If T invariance is violated as well, each of the constants becomes complex. It was up to experiment to identify which variants are realized in nature and what the relations are between the constants.

Analyzing the experiments on the β-decay of nuclei in their totality, *Feynmann, Gell-Mann, Marshak,* and *Sudarshan* [122] concluded that the V and A variants of the weak interaction are responsible for the β-decay of nuclei and for the decays of elementary particles. Experiments with polarized neutrons proved for the first time that the C_V and C_A constants have opposite signs; i.e., the V–A variant of the weak interaction is realized. According to the hypothesis of [122], the weak interaction is universal; i.e., it is characterized for any pair of particles by the same V–A variant and by the same coupling constant G, the so-called universal weak coupling constant (Fermi constant). The V–A structure of the weak charged current reflects the nonconservation of P and C parities because the transformations of the V and A terms differ in sign, both under P inversion and under C conjugation. As a result, the terms $VV^+ + AA^+$ in the Lagrangian of the type $(V-A)(V-A)^+$ retain the sign, while the terms $VA^+ + AV^+$ reverse it (the cross superscript denotes Hermitian conjugation).

6.5 The Structure of the Nucleon-Nucleon Weak Interaction

According to the Weinberg-Salam standard model of the electroweak interaction [82], the Hamiltonian of the nucleon-nucleon weak interaction is written as the product of charged (J^+) and neutral (J^z) currents. The charged hadronic current contains components of the quark multiplet $\psi = (\bar{u}, \bar{d}, \bar{s}, \bar{c})$ of left-hand (L) helicity [124]:

$$J^+ = \cos\theta_C (\bar{d}u)_L + \sin\theta_C (\bar{s}u)_L + \cos\theta_C (\bar{s}c)_L - \sin\theta_C (\bar{d}c)_L . \quad (6.3)$$

The first two terms of (6.3) correspond to the charged current of the Cabibbo model [125], and $\theta_C \simeq 13°$ is the Cabibbo angle. The u and d quarks form an isotopic doublet ($T = 1/2$), while the s and c quarks are isoscalar ($T = 0$); consequently, the current $\bar{d}u$ is isovector ($\Delta T = 1$), while the current $\bar{s}u$ is isospinor ($\Delta T = 1/2$). Each of the currents $\bar{d}u$ and $\bar{s}u$ consists of a sum of the polar and axial currents (V and A).

The neutral hadronic current is written in the form

$$J^z = J_0 - 2\sin^2\theta_W J^\gamma , \quad \text{where} \quad (6.4)$$

$$J_0 = \tfrac{1}{2}(\bar{u}u - \bar{d}d + \bar{c}c - \bar{s}s)_L \quad (6.5)$$

is the neutral component of the V–A isotriplet,

$$J^\gamma = |\tfrac{1}{2}(\bar{u}u - \bar{d}d) + \tfrac{1}{6}(\bar{u}u + \bar{d}d) - \tfrac{1}{3}\bar{s}s + \tfrac{2}{3}\bar{c}c|_{L+R} \tag{6.6}$$

is the electromagnetic (vector) current, $\theta_W \approx 30°$ is the Weinberg angle, and R is the subscript for right-hand helicity.

Nucleons (as well as the π and ϱ mesons) contain only u and d quarks, so that the Hamiltonian of the nucleon-nucleon weak interaction taking into account only charged currents $H_W = (G/\sqrt{2})J^+J^+$ contains both isoscalar ($\Delta T = 0$) and isotensor ($\Delta T = 2$) terms:

$$H_{0,2} \propto \cos^2\theta_C (\bar{d}u)_L (\bar{u}d)_L \;,$$

proportional to $\cos^2\theta_C$. If neutral currents are also taken into account, the product of the terms $\bar{u}u - \bar{d}d$ (isovector) and $\bar{u}u + \bar{d}d$ (isoscalar) gives an isovector component ($\Delta T = 1$) to the term of the Hamiltonian proportional to $\cos^2\theta_C$.

In order to calculate the effects of parity violation in nuclei, it is convenient to use as a first approximation the effective parity-violating one-particle potential. It can be interpreted as the mean potential acting on one nucleon beyond the core of the nucleus and applied by the remaining nucleons. This one-particle potential of the weak interaction of a nucleon in a nucleus is written in the form [126]:

$$V_{PV} \approx G s p \varrho / 2m \;, \tag{6.7}$$

where s, p, and m are the spin, momentum, and mass of the nucleon, respectively, and ϱ is the density of nucleons in the nucleus (in the system of units in which $\hbar = c = 1$).

In order to evaluate the amplitude of mixing of the states with opposite parities in the one-particle approximation, we divide V_{PV} by the characteristic energy of a nucleon in the nucleus, $\omega \approx p^2/2m$. Taking into account that in a nucleus, $p \approx m_\pi$, $\varrho \approx m_\pi^3$, we arrive at an estimate of the relative value of the nucleon-nucleon weak coupling, in the form of a dimensionless parameter

$$F \approx V_{VP}/\omega \approx Gm_\pi^2 = (10^{-5}/m^2)m_\pi^2 \approx 3\times 10^{-7} \;. \tag{6.8}$$

The parameter F being very small, it was necessary to adjust the experimental conditions in the first experiments on the nucleon-nucleon weak interaction so as to increase the initial value of F by enhancing the nucleon-nucleon weak interaction.

6.6 Enhancement Mechanisms of the Nucleon-Nucleon Weak Interaction

A classification of the enhancement mechanisms for P-odd effects in nuclear intractions was given by *Shapiro* [127]. The main enhancement mechanism in experiments with polarized neutrons is the so-called dynamic enhancement related to the high-level density in the compound nucleus. Let us consider, without going into unnecessary details, the factors causing this enhancement.

It is justifiable to treat the nucleon-nucleon weak interaction as a perturbation in the system of strongly interacting particles. The total interaction Hamiltonian of a nuclear system can then be written as a sum,

$$\mathcal{H} = H_0 + H_w , \tag{6.9}$$

where H_0 is the main, scalar, part of the Hamiltonian, and H_w is the pseudoscalar parity-nonconserving component which describes the interaction mixing the states of the nuclear system with opposing parities. In the first approximation of perturbation theory, the eigenstates of the operator $\mathcal{H}$ are determined by the wave functions

$$\Psi_i = \psi_i + \sum_{j \neq i} \frac{\langle j|H_w|i\rangle}{E_j - E_i} \psi_j , \tag{6.10}$$

where ψ_i and ψ_j are the eigenfunctions of the nonperturbed Hamiltonian H_0, having parities of opposite signs; E_i and E_j are the energies of the corresponding nonperturbed states. The sum is taken over all states whose parity is the opposite of the parity of the state ψ_i. The expression (6.10) is valid if the condition

$$\langle j|H_w|i\rangle \ll |E_j - E_i|$$

holds, as it does in the case under consideration. Assuming that the sum (6.10) is dominated by the contribution of the level closest to that with energy E_i, we can simplify the expression (6.10)

$$\Psi_i = \psi_i + \alpha \psi_j , \quad \text{where} \tag{6.11}$$

$$\alpha = \langle j|H_w|i\rangle/(E_j - E_i) \tag{6.12}$$

is the mixing coefficient.

For ground states of nuclei, the value of α is close to that of F. However, for highly excited states of nuclei, α may grow substantially because the levels with energies E_i and E_j may be quite close, and hence, the denominator in (6.12) may be very small. A correct estimation of α was first made by *Blin-Stoyle* [128]. We can write it in the form [126]:

$$\alpha = \langle M \rangle / D , \quad \text{where} \tag{6.13}$$

$$\langle M \rangle = \langle j|H_w|i\rangle$$

is the matrix element of the one-particle operator H_w, and D is the mean spacing between the levels of the compound nucleus. Estimates show that

$$\langle M \rangle \approx F\sqrt{D\Delta E},$$

where ΔE is the mean spacing in energy between one-particle states. Hence, the mixing coefficient is $\alpha \approx F\sqrt{\Delta E/D}$. The quantity

$$R_{\text{dyn}} = \sqrt{\Delta E/D} \tag{6.14}$$

was termed the dynamic enhancement factor.

Let us now consider the mechanism of the so-called kinematic enhancement which arises from the interference of either input or output channels of the neutron capture reaction producing a compound nucleus. The situation can be analyzed by using Feynmann diagrams for the reaction amplitudes (Fig. 6.2) [129].

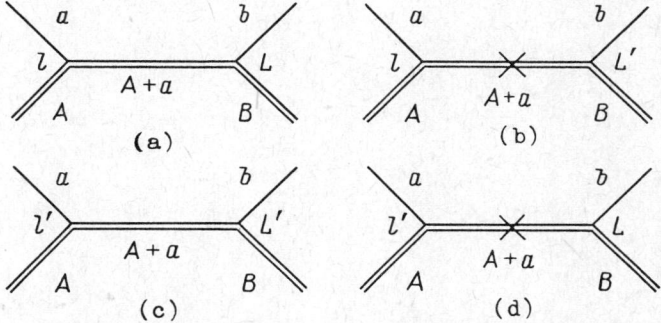

Fig. 6.2. Feynman diagrams for reaction amplitudes: *(A)* initial nucleus; *(B)* final nucleus; *(a)* capture neutron; *(A+a)* compound nucleus; *(b)* particle emitted from the nucleus; (l, l') orbital momenta of the ingoing neutron a and of the nucleus A; (L, L') orbital momenta of the outgoing particle b and nucleus B. The weak nucleon-nucleon interaction is marked by a cross

The interference of amplitudes shown in Figs. 6.2a and d produces an interference term proportional to $\sqrt{\Gamma_{l'}^a/\Gamma_l^a}$, where $\Gamma_{l'}^a$ and Γ_l^a are the widths of the corresponding input channels. This term describes the interference of input channels for the same output channel. The interference of amplitudes shown in Figs. 6.2a and b produces an interference term proportional to $\sqrt{\Gamma_{L'}^b/\Gamma_L^b}$, where $\Gamma_{L'}^b$ and Γ_L^b are the widths of the corresponding output channels. This term describes the interference of output channels for the same input channel. This mechanism will be described in more detail in Sect. 8.1.1.

The last enhancement mechanism discussed in [127], the so-called structural mechanism, operates when the matrix element of a regular transition is suppressed in view of the selection rules for the additional quantum numbers, while the matrix element of a nonregular transition is either not suppressed or suppressed to a lower degree. Such cases are more typical for deformed nuclei. This mechanism is important for the emission of circularly polarized gamma radiation from nonpolarized nuclei. Since this effect is not related to polarized neutrons, it will not be considered in this book.

7. Decay of Polarized Neutrons

The study of decays of polarized neutrons provided one of the first experimental proofs of the Lee and Yang hypothesis on nonconservation of P parity in the weak interaction [118]. The neutron is the most suitable object for studies of this kind because it constitutes the simplest nucleus whose β-decay follows the scheme

$$n \to p + e^- + \tilde{\nu}_e \ .$$

7.1 Fundamentals of the Experimental Method

In their analysis of the decay of polarized neutrons, *Burgy* et al. [130] employed the following correlation formula for the probability of the decay of a free neutron accompanied by the emission of an electron with energy E_e and momentum $\boldsymbol{p}_e$ and an antineutrino with energy $E_{\tilde{\nu}}$ and momentum $\boldsymbol{p}_{\tilde{\nu}}$:

$$w = N(E)\xi\left\{1 + a\frac{\boldsymbol{p}_e \boldsymbol{p}_{\tilde{\nu}}}{E_e E_{\tilde{\nu}}} + A\frac{\langle\boldsymbol{s}\rangle\boldsymbol{p}_e}{sE_e} + B\frac{\langle\boldsymbol{s}\rangle\boldsymbol{p}_{\tilde{\nu}}}{sE_{\tilde{\nu}}} + D\frac{\langle\boldsymbol{s}\rangle(\boldsymbol{p}_e\times\boldsymbol{p}_{\tilde{\nu}})}{sE_e E_{\tilde{\nu}}}\right\} \ , \quad (7.1)$$

where $N(E)$ is a function of the shape of the electron spectrum, $\langle\boldsymbol{s}\rangle$ denotes the mean value of neutron spin projections on the specified axis, and ξ, a, A, B, D are correlation coefficients which depend on the constants of weak interaction variants and on the matrix elements (see Sect. 6.4). Using this dependence, it is possible to establish experimentally which variants of interaction do take place in nature and what the relations are between the constants.

Let us discuss the physical meaning of individual terms in the sum (7.1). The contribution of each term can be separated from those of other terms. The term containing the product $\boldsymbol{p}_e \boldsymbol{p}_{\tilde{\nu}}$ describes the electron-neutrino correlation. It is independent of neutron polarization and thus need not be considered here. The experimental data on this correlation are summarized in the review [131] and in the monograph [75].

The remaining three terms depend on the neutron polarization $P_n = \langle\boldsymbol{s}\rangle/s$ and can be measured in the decay of polarized neutrons. The term with the product $\langle\boldsymbol{s}\rangle\boldsymbol{p}_e$ describes the angular distribution of electrons with respect to the direction of the neutron spin. The next term, containing the product $\langle\boldsymbol{s}\rangle\boldsymbol{p}_{\tilde{\nu}}$, characterizes the angular distribution of antineutrinos with respect to the direction of the neutron spin. These two products are pseudoscalars (see Sect. 6.1), and hence, the appearance of any one of them in the neutron decay points to the nonconservation of P parity. The last term in the expression (7.1) is the neutron-electron-antineutrino correlation. This term reverses its sign under time reversal (see Table. 6.1). Therefore, if this term were observed, the neutron decay would not be invariant with respect to time reversal. The violation of time reversal invariance is taken into account by introducing complex constants of the weak interaction variants. The ratio of the constants C_A and C_V then takes the following form:

$$\lambda = C_A/C_V = |\lambda| \exp(i\varphi) . \tag{7.2}$$

If the process is invariant under time reversal, then $\varphi = \pi$.

7.2 Spatial Parity Violation

7.2.1 Experiment

The first measurements of the coefficients A and B in (7.1) were carried out by *Novey* and his coworkers [130, 132], and *Clark* and *Robson* [133]. The most accurate values of these coefficients were obtained with the apparatuses at Argonne National Laboratory (ANL) [53, 134] and at the Kurchatov Institute of Atomic Energy (IAE) [135, 136].

Let us consider these studies. The main difficulty in measuring the coefficient B with the antineutrino-neutron spin correlation arises because the antineutrino is not directly recorded in the experiments, and its momentum can be evaluated only by taking into account the directions of momenta and the energies of the electron and the proton created in the decay. When the coefficient A is measured in the electron-neutron spin correlation, one need not, in principle, know the direction of emission of the antineutrino. However, the problems of background reduction require the introduction of an electron detector whose pulses are recorded in coincidence with the pulses of the proton detector. Therefore, any apparatus for measuring the coefficients A and B includes two detectors, one for electrons and the other for protons, in a delayed coincidence circuit. The upper bound energy of the electron spectrum is 782 keV, so that the electrons are detected by scintillation counters. The maximum energy of decay protons is about 800 eV, and they are detected either by electron multipliers or by special scintillation counters; the proton energy is deduced from the time of flight across a certain distance.

Figure 7.1 depicts the ANL apparatus. The proton detector is a wide-window electron multiplier mounted in the vacuum chamber and facing the electron detector (scintillating plastic counter). The neutron decay region was the zero-field space between the grids, in which protons moved at their initial velocities. If an antineutrino moved toward the electron detector, the velocities

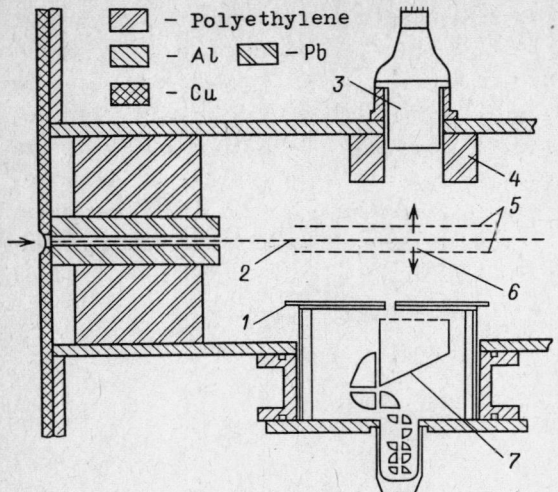

Fig. 7.1. The ANL apparatus for studying polarized neutron decay [53]: *(1)* beryllium anode ($V = 0$), *(2)* neutron beam, *(3)* electron detector (scintillation plastic), *(4)* shielding of the detector, *(5)* grids ($V = 9000\,V$), *(6)* two variants of the magnetic field direction, *(7)* proton detector cathode ($V = -1135\,V$)

of recoil protons were greater than when the antineutrino was emitted in the direction of the proton detector. The direction of antineutrino emission was thus a function of the neutron time of flight across the zero-field gap. This dependence was found quantitatively in computer calculations. Neutron spins were directed once toward the electron detector and then away from it. The coefficient A was determined by measuring the difference between coincidence counts for the parallel and antiparallel directions of s with respect to p_e. The coefficient B was found from the shift of temporal proton pulse distributions for the opposite orientations of neutron spins.

Figure 7.2 illustrates the IAE apparatus for measuring the coefficient A. A scintillation counter acted as the electron detector. A special scintillation counter with a very thin layer of CsI(Tl) was developed for proton detection. The change in coincidence count due to spin reversal depends not only on A but

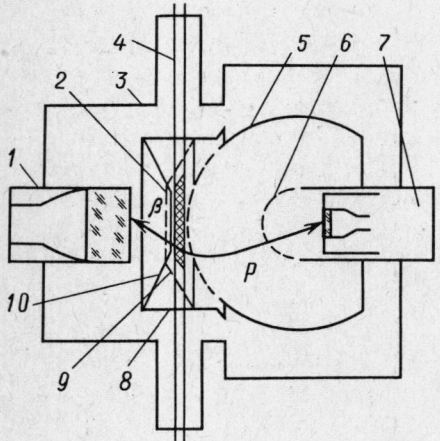

Fig. 7.2. The IAE apparatus for studying polarized neutron decay [135]: *(1)* electron detector (scintillation plastic and photomultiplier), *(2)* grid, *(3)* vacuum chamber, *(4)* neutron beam, *(5)* spherical electrode with grid, *(6)* smaller spherical grid, *(7)* proton detector, *(8)* screening electrode, *(9)* conical grid, *(10)* diaphragm defining the working part of the beam

also on the large coefficient B (high antineutrino-neutron spin correlation). It was shown [137] that the effect of one correlation on the other can be eliminated if the region of the beam from which decay electrons are detected is limited by a diaphragm. In the proton collection system, however, all recoil protons are recorded from each point of this region, regardless of the direction of emission of antineutrinos.

The coefficient B was measured at the Institute of Atomic Energy by an apparatus similar to that for measuring the three-vector correlation (the coefficient D), as described in Sect. 7.3. Additional details concerning the IAE instruments can be found in the review by *Erozolimsky* [131].

7.2.2 Results

Table 7.1 lists the most accurate data on the coefficients A and B reported by a number of laboratories, the averaged values, and the theoretical predictions for these coefficients calculated under the following assumptions:

1. The β-decay of neutrons is described by the axial-vector variant of the weak interaction with the constant C_A and the vector variant of the weak interaction with the constant C_V, the sign of C_A/C_V being negative (The V–A variant);
2. The antineutrinos are completely polarized in the direction of momentum, as required by the two-component neutrino theory (the antineutrino has right-hand helicity) [138].

Table 7.1. The values of the coefficients A and B

Coefficient	Experiment	Source	Average value [75]	Theory
A	−0.113±0.006 −0.114±0.005	ANL [134] IAE [136]	−0.1136±0.0038	−0.1
B	1.00 ±0.05 0.995±0.034	ANL [53] IAE [135]	0.998 ±0.026	1.0

The experimental values of A and B are clearly in good agreement with the theoretical estimates under the assumptions stemming from the V–A variant of the weak interaction.

No agreement is found between the experimental data and the theoretical predictions if other possible combinations of weak interaction variants (V+A, S+T, S−T) are used, or if another helicity is assigned to the antineutrino. The following conclusions were drawn from the experiments with polarized neutron decay:

1. Spatial parity is not conserved in this process;

2. The vector and axial-vector variants of the weak interaction predominate, the contribution of the scalar variant being less than 45%, and that of the tensor variant less than 14%;
3. The ratio of the constants of weak interaction variants is
$\lambda = -1.254 \pm 0.006$ [139];
4. The antineutrino has right-handed helicity, which gives $C'_V = C_V$ and $C'_A = C_A$; hence, the spatial (P) parity is completely broken.

These conclusions are borne out by other β-decay experiments and are generalized in the universal four-fermion weak interaction theory [122] (see Sect. 6.4).

The conclusion on the total longitudinal polarization of the antineutrino (or the neutrino) involves the assumption that the antineutrino mass is identically zero. It has been reported recently that the antineutrino may have a very small mass [140]. This result cannot significantly alter the conclusions listed above.

Noted added in proof: Recently the experimenters at ILL [P. Bopp, D. Dubbers, L. Hornig, E. Klemt, J. Last, H. Schütze, S.J. Freedman, O. Schärpf: Phys. Lett. **56**, 919 (1986)] have achieved a record accuracy in measuring the coefficient A in the decay of polarized neutrons. They obtained $A = -0.1146 \pm 0.0019$, which gives $\lambda = -1.262 \pm 0.005$.

7.3 Investigation of Invariance Under Time Reversal

7.3.1 Experiment

In order to analyze the invariance of the decay of polarized neutrons under time reversal, one has to measure the coefficient D in the formula (7.1) in front of the last term which describes the neutron-electron-antineutrino correlation, because this term reverses its sign under time reversal operation. This correlation involves three noncomplanar vectors: neutron spin s, electron momentum p_e, and antineutrino momentum $p_{\bar{\nu}}$. The corresponding experiment is frequently referred to, accordingly, as the three-vector experiment.

The idea behind the experiment is demonstrated by Fig. 7.3. The directions of the neutron spin and the momenta of decay products are reversed for the observer, using "reversed" time. Figure 7.3b shows the vector arrangement corresponding to reversed time. A 180° rotation of the electron detector and the imaginary antineutrino detector around the neutron spin direction (this is equivalent to the rotation of the reference frame) restores the original arrangement (as in Fig. 7.3a). Only the orientation of the neutron spin is thereby reversed. Obviously, these arguments assume that space is isotropic.

We can conclude, therefore, that the reversal of neutron spins is equivalent to time reversal. If the neutron decay rate with "reversed" vectors is equal to the neutron decay rate with "nonreversed" vectors, the decay process is symmetrical with respect to time reversal; i.e., it is invariant under time reversal.

The invariance of the decay of polarized neutrons under time reversal was first studied by *Robson* [133] and *Novey* [132] with their respective coworkers.

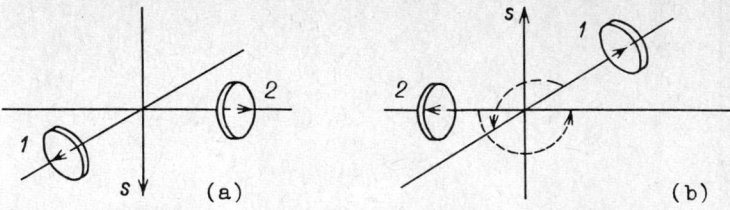

Fig. 7.3. Principle of the experiment searching for noninvariance with respect to time reversal in polarized neutron decay: *(1)* antineutrino detector, *(2)* electron detector

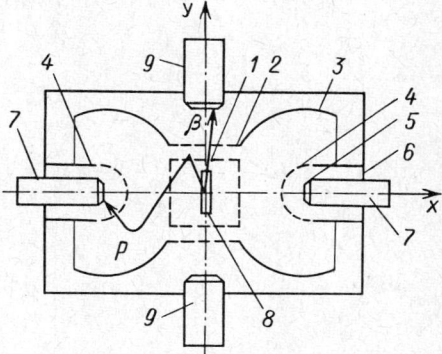

Fig. 7.4. The IAE apparatus for studying the invariance of polarized neutron decay with respect to time reversal [141]: *(1)* inner grid cylinder ($V = 24.2\,\mathrm{kV}$), *(2)* outer grid cylinder, *(3)* outer hemisphere ($V = 25\,\mathrm{kV}$), *(4)* grid hemisphere, *(5)* CsI(Tl) crystal, *(6)* chamber, *(7)* proton counter, *(8)* cross section of the neutron beam, *(9)* electron detector

More accurate experiments on measuring the coefficient D were carried out at the IAE [141] (Fig. 7.4). The neutron spins were oriented along the beam (perpendicularly to the plane of the figure), so that the directions p_e, $p_{\bar{\nu}}$, and s formed a triad of mutually orthogonal vectors. An important feature of the apparatus was its symmetry (two pairs of identical detectors). The same detectors that served to measure A and B were used in the measurements of D (see Sect. 7.2). The system recorded four groups of delayed coincidences for two opposite directions of the neutron spins. The spins were flipped every 1000 s.

This experiment implemented a unique approach to determining the direction of antineutrino emission. While in [132] this was done by selecting neutron decay events with a prescribed direction of proton ejection, in [141], proton pulses were analyzed for time of flight across the zero-field gap for a fixed direction of electron emission. It was then possible to calculate the angle of antineutrino emission, resorting directly to the momentum conservation law [142].

Further progress in the analysis of the invariance of polarized neutron decay under time reversal became possible as more intensive polarized neutron beams were obtained. A vertical channel of polarized neutrons was constructed for this express purpose at the IAE reactor (see entry 7 of Table 1.1); this channel led to a higher accuracy of measuring the coefficient D [143]. A similar experiment was carried out at the high-flux ILL reactor at Grenoble with a beam of cold polarized neutrons [144].

Still greater attention was paid in a later IAE work [145] to the symmetry of the system and to the eliminating of the deviation of the neutron polarization vector from the longitudinal axis of the apparatus. To achieve this, the system was designed to allow a 90° rotation of the entire proton-focusing and -detection unit around the axis passing through the proton detectors. This study revealed effects caused by the nonuniform distribution of the directions of the neutron polarization vectors across the beam cross section; this nonuniformity could produce a spurious asymmetry in the coincidence count. The experiment was run so as to reverse the sign of the spurious asymmetry with respect to the sought asymmetry, by peridocally varying the experimental conditions. The spurious effect was thereby eliminated.

7.3.2 Results

The data of the most accurate experiments measuring D are shown in Table 7.2. These results demonstrate the absence of triple correlation in the neutron decay, and hence, confirm that this process is invariant under time reversal. The best value obtained for the paramter φ [see (7.2)] [139] is

$$\varphi = (180.11 \pm 0.17)° \ .$$

The experiments measuring angular correlations in polarized neutron decay have thus corroborated the hypothesis of spatial parity nonconservation in the weak interaction, established the upper bound on the violation of invariance under time reversal, and provided a solid foundation for constructing the V–A variant of weak interaction theory.

Table 7.2. Results of the measurements of the coefficient D

D_{exp}, $[10^{-3}]$	Source	$\overline{D}$, $[10^{-3}]$ [139]
−2.7±3.3	IAE [143]	
−1.1±1.7	ILL [144]	−0.7±1.4
2.2±3.0	IAE [145]	

8. Anisotropy of Gamma Rays Emitted by Nuclei After Polarized Neutron Capture

The study of the P-odd anisotropy of the gamma radiation emitted by nuclei after polarized neutron caputre constituted the first experiment to detect the violation of spatial parity in nuclear interactions [146].

The search for a nucleon-nucleon interaction which would not conserve P parity was initiated soon after the discovery of P parity violation in the β-decay. The theoretical foundation of the search was the hypothesis of the universal four-fermion interaction (see Chap. 6).

Wilkinson [147] was the first to consider experimental possibilities of detecting a nucleon-nucleon potential not conserving P parity. These possibilities can be classified into two groups. The first group is comprised of the experiments searching for the violation of the absolute selection rules. This primarily involves the search for parity-forbidden α-decays of the type $J^\pi \to 0^{\pi'}$, where $\pi' = (-1)^{J+1}\pi$ (J is the nuclear spin, and π, π' are the parities of the states before and after the α-decay, respectively). The probability of observing this decay in such experiments is proportional to F^2, where F is a dimensionless parameter charcterizing the relative strength of the weak interaction between nucleons (see Sect. 6.5). The second group is comprised of experiments manifesting the interference of nuclear states with opposing parities. Experiments of this kind are designed to measure the anisotropy in the emission of gamma quanta by polarized nuclei, the circular polarization of gamma quanta emitted by nonpolarized nuclei, or the anisotropy in the emission of alpha particles and fission fragments by polarized nuclei. In this group of experiments, the emission asymmetry is proportional to F; i.e., it is larger by many orders of magnitude than the probability in experiments of the first group.

In the present chapter, we discuss only the experiments designed to detect an asymmetry in the gamma emission from nuclei polarized through the capture of polarized neutrons. When comparing the experimental results among themselves and with theoretical predictions in Sect. 8.2, we mention the results of other experiments which confirmed the violation of spatial parity in nuclear interactions. The subsequent chapters describe some new experimental approaches to analyzing P parity violation in nuclear interactions; for instance, the anisotropy of α-emission, the anisotropy of fission products, and the coherent parity violation, studied with polarized neutron beams.

Unfortunately, it would be impossible to compare the results of all relevant experimental studies of P parity violation in nuclear interactions within the space of the present book. We suggest that the reader turn to the review papers [124, 148, 149] for more detailed comparative data.

Finally, Sect. 8.3 discusses polarized-neutron experiments studying the invariance under time reversal.

8.1 Fundamentals of the Experimental Method

8.1.1 Estimate of the Expected Enhancement of P-Odd Effects

Two mechanisms of enhancement of P-odd effects are essential for the emission of gamma quanta by nuclei after polarized neutron capture: the dynamic and the kinematic mechanisms (see Sect. 6.6). The estimates below are given for medium-mass nuclei (cadmium and tin) because the experiments work chiefly with such nuclei.

The average spacing between one-particle states of nuclei with $A \approx 100$ and excitation energy $E \approx 10$ MeV is $\Delta E \approx 1$ MeV, while the average spacing between the levels of a compound nucleus is $D \approx 10-100$ eV. Consequently, the dynamic enhancement factor $R_{\text{dyn}} = \sqrt{\Delta E/D}$ [see (6.14)] may reach 100.

In order to evaluate the kinematic factor, we recall that the (n, γ)-reaction with polarized neutrons is accompanied by the interference of output channels for the same input channel; namely, the interference of the regular (ML) and nonregular $(\widetilde{EL})$ electromagnetic nuclear transitions. In this context, an electromagnetic transition is said to be regular if it is not mediated by the weak interaction, and nonregular if it manifests the nucleon-nucleon weak interaction. We recall that ML is a symbol for a magnetic transition with multipolarity L, and EL symbolizes an electric transition of the same multipolarity L. A nonregular transition is denoted by a tilde over the transition symbol.

If a compound nucleus $A + a$ (see Fig. 6.2) is polarized via the capturing of a polarized neutron a by a nucleus A, an asymmetry is produced in the emission of a gamma quantum (particle b) relative to the direction of polarization. In this case, the interference term is proportional to the quantity $R_{\text{kin}} = \sqrt{\Gamma(\widetilde{EL})/\Gamma(ML)} = |\widetilde{EL}|/|ML|$, where $|\widetilde{EL}|$ and $|ML|$ are the matrix elements of the corresponding transitions.

According to one-particle estimates, $|\widetilde{EL}| \propto kr$, where k is the wave number, r is the radius of the nucleus, and $|ML| \propto (v/c)kr$, where v is the velocity of nucleons in the nucleus. Thus, in the present case, $R_{\text{kin}} \approx c/v \approx 10$.

We can therefore expect that for some nuclei, the enhancement factor reaches $R = R_{\text{dyn}} R_{\text{kin}} \approx 10^3$.

8.1.2 Angular Distribution of Gamma Quanta Emitted by Nuclei After the Capture of Polarized Thermal Neutrons

This angular distribution of gamma quanta can be written in the form

$$w(\theta) = \text{const}(1 + P_n a \cos \theta) , \qquad (8.1)$$

where a is the coefficient of the P-odd asymmetry; as follows from (8.1)

$$a = [w(0) - w(180°)]/P_n[w(0) + w(180°)] ,\qquad(8.2)$$

and θ is the angle between the direction of neutron beam polarization and the gamma-quantum momentum. Here, $w(0)$ and $w(180°)$ are two quantities proportional to the numbers of pulses recorded by a detector, which is insensitive to the polarization of gamma rays, whenever the angles between the neutron beam polarization direction and the gamma quantum momentum are equal to 0 and 180°, respectively. Sometimes a quantity twice as large is used: $a' = 2a$.

The general theory of the angular distribution of gamma emission can be found in the review [150] and in the monographs [102, 151]. We conclude this section with the formulas for the asymmetry coefficient, derived by *Blin-Stoyle* [148] on the basis of the general theory of angular correlations; they will be necessary for the interpretation of the experiments studying P-odd effects.

First we consider the case in which a regular γ-transition is a mixed $E(L+1) - ML$ transition. Then the nonregular transition may be an $\widetilde{EL}$ transition [we neglect the admixture of the $\widetilde{M(L+1)}$ transition]. In this case, the asymmetry coefficient is

$$a = \frac{2B_1(J_c)}{1+\delta^2}\left[F_1(LLJ_fJ_c)\frac{|EL|}{|ML|} + \delta^2 F_1(LL+1J_fJ_c)\frac{|EL|}{|E(L+1)|}\right],\qquad(8.3)$$

where

$$B_1(J_c) = [3/4 + J_c(J_c+1) - J_i(J_i+1)]/[3J_c(J_c+1)]^{1/2}\qquad(8.4)$$

is proportional to the orientation parameter of a nucleus after the capture of a polarized s-neutron; δ is the mixing parameter of the gamma transition, described by the formula

$$\delta = \langle J_f|L+1|J_c\rangle/\langle J_f|L|J_c\rangle .\qquad(8.5)$$

The coefficients $F_1(LLJ_fJ_c)$, having a simple analytical form [152]

$$F_1(LLJ_fJ_c) = -\frac{\sqrt{3}}{2}\frac{L(L+1) + J_c(J_c+1) - J_f(J_f+1)}{L(L+1)[J_c(J_c+1)]^{1/2}} ,\qquad(8.6)$$

are tabulated in [150]; J_c, J_f, J_i are the spins of the compound nucleus emitting a gamma quantum, of the final and the initial nuclei, respectively.

If the regular transition is a pure ML transition and the nonregular one is an $\widetilde{EL}$ transition, then we assume $\delta = 0$, substitute (8.4) into (8.3), and arrive at the asymmetry coefficient

$$a = 2\frac{3/4 + J_c(J_c+1) - J_i(J_i+1)}{[3J_c(J_c+1)]^{1/2}}F_1(LLJ_fJ_c)\frac{|EL|}{|ML|} = 2ARF ,\qquad(8.7)$$

where

$$A = \frac{3/4 + J_c(J_c+1) - J_i(J_i+1)}{[3J_c(J_c+1)]^{1/2}}F_1(LLJ_fJ_c)\qquad(8.8)$$

and $RF = |EL|/|ML|$ is the ratio of matrtix elements.

The factors singled out in the expression for the asymmery coefficient a are: A, which depends only on the characteristics of the nucleus (the spins

J_i, J_c, and J_f) and of the gamma transition (its multipolarity L), and the factor RF, which gives the relative contribution F of the nucleon-nucleon weak interaction, times the enhancement factor $R = R_{\text{dyn}} R_{\text{kin}}$.

8.2 Experimental Investigation of Spatial Parity Violation in Nuclear Interactions

8.2.1 Choice of Nuclei

Cadmium nuclei are a very convenient object in the search for the P-odd asymmetry in (n,γ) reactions. Actually, only one Cd isotope, ^{113}Cd, with $J_i^\pi = 1/2^+$, contributes aprreciably to the (n,γ) reaction. The thermal neutron capture has a cross section $\sigma = 2520 \times 10^{-24}\,\text{cm}^2$; it is caused by an s-resonance with the spin $J_c = 1$, at an energy of 0.178 eV, resulting in the formation of an excited state of ^{114}Cd with $J_c^\pi = 1^+$. Figure 8.1 shows that part of the formation and decay scheme of this nucleus which is relevant in this context.

The transition from the compound state $J_c^\pi = 1^+$ to the ground state characterized by $J_f^\pi = 0^+$ is an $M1$ transition with an energy of 9.04 MeV. The state density in cadmium at the excitation energy of about 9 MeV is high enough to suggest that levels with identical spins but opposite parities should lie close to each other. As mentioned in Sect. 8.1, this situation leads to a dynamic enhancement of the P-odd effect. Furthermore, since the $1^+ \to 0^+$ transition is an $M1$ transition, the nonregular transition must be $\widetilde{E1}$. The $M1 - \widetilde{E1}$ interference results in a kinematic enhancement of the effect.

The reaction $^{117}\text{Sn}(n,\gamma)^{118}\text{Sn}$ has similar characteristics, but tin specimens for this kind of work must be enriched in ^{117}Sn.

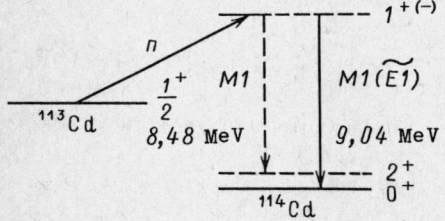

Fig. 8.1. Part of the scheme of formation and decay of the ^{114}Cd nucleus

8.2.2 Specifics of the Reactions $^{113}\text{Cd}(\vec{n},\gamma_0)$ and $^{117}\text{Sn}(\vec{n},\gamma_0)$

The spin factor A found from the expression (8.8) is equal to -1 for the considered $1^+ \to 0^+$ transitions. Estimates of the enhancement factor for cadmium gave $R \approx 10^3$ [128]. Since $F \approx 10^{-7}$ (see Sect. 6.5), the asymmetry coefficient is expected to be of the order of 10^{-4}. The closest in energy transition from the state 1^+ is the transition to the first excited state 2^+ for both nuclei. This is an $M1$ transition. If the initial state 1^+ contains a P-odd admixture (see Sect. 6.6), then the gamma quanta of the $1^+ \to 2^+$ transition necessarily have anisotropic angular distribution. Unfortunately, the sign of the spin factor A

is then reversed: $A = +0.5$; hence, the sign of the asymmetry coefficient is opposite to that for the main transition $1^+ \to 0^+$. A decrease in the coefficient A is compensated for by the fact that the $1^+ \to 2^+$ transition is more intensive than the $1^+ \to 0^+$ transition; consequently, if these two transitions are poorly resolved in the apparatus because of inadequate energy resolution, the effects with opposite signs may cancel each other out. This is especially true for the ^{114}Cd nucleus, because it has a smaller difference between the energies of the competing gamma transitions (only 0.5 MeV) than the ^{118}Sn nucleus does.

8.2.3 First Attempt at Detecting a P-Odd Effect in (n, γ) Reactions

Originally, the search for a P-odd asymmetry in the emission of gamma quanta was undertaken in [153], where the asymmetry was measured in the emission of gamma quanta along and against the direction of polarization of the neutron beam. Taking into account the finite sizes of gamma-ray detectors and of the target, we rewrite the expression (8.1) in the form

$$N_\pm = \text{const}(1 \pm P_n a \Omega) \, , \tag{8.9}$$

where $N_\pm$ is the number of counts of the gamma-ray detector for the case in which the γ-quantum momentum and the neutron spin are parallel (the plus subscript) or antiparallel (the minus subscript), and $\Omega = \overline{\cos \theta}$ is a geometric factor taking into account the finite sizes of the detector and the target.

The asymmetry coefficient a can be found by calculating the difference or the ratio of N_+ and N_-. In order to diminish the effects of apparatus asymmetry and electronic drift, it is desirable to have two simultaneously operating, practically identical channels, installed on two sides of the target and recording gamma quanta with momenta along and against the beam polarization direction, respectively.

Haas et al. [153] studied the angular distribution of gamma quanta emitted by nuclei of cadmium, indium, and silver after polarized neutron capture. A beam of polarized neutrons at an energy of 0.09 eV, polarized to about 80%, was obtained by the reflection of the beam from (111) planes of a magnetized cobalt-iron alloy single crystal. The neutron flux at the target (over the total specimen area of about 6.5 cm^2) was 2×10^4 neutron/s. Gamma quanta emitted along and against the neutron beam polarization direction were recorded by two identical scintillation spectrometers with NaI(Tl) crystals. The pulses from photomultipliers were amplified and fed to single-channel differential analyzers. The polarization direction of the neutron beam was periodically rotated by 180° (the period was fairly long).

No asymmetry was detected for the $1^+ \to 0^+$ transition in the ^{114}Cd nucleus (9.04 MeV transition energy):

$$a = (1.2 \pm 7.8) \times 10^{-4} \, .$$

The coefficient a for silver and indium was measured at a still lower accuracy.

The authors of [153] used a highly overestimated enhancement factor R for the reaction $^{113}\text{Cd}(\vec{n},\gamma_0)$ and derived far-reaching conclusions on the smallness of F. This mistake was pointed out by *Blin-Stoyle* [128]. With more realistic values of R, the results of [153] yield $F \leq 10^{-6}$.

8.2.4 Experimental Discovery of the Effect

The polarized neutron beam used at the Institute of Theoretical and Experimental Physics (ITEP) in Moscow was considerably more intensive (see entries 1 and 3 in Table 1.1), so that the accuracy achieved in [153] was improved by an order of magnitude. According to the estimates given above (see Sect. 8.1), there was hope that a P-odd effect in the $^{113}\text{Cd}(\vec{n},\gamma_0)^{114}\text{Cd}$ reaction would be detectable. The experiment was repeated three times with modified instruments [7, 19, 146]. The polarized neutron beams for all three ITEP experiments were generated in the horizontal channel of the heavy-water reactor by reflecting from magnetized cobalt mirrors (Fig. 8.2). Having passed through a number of collimators and magnetic guides, the polarized beam struck a metallic cadmium target, 0.4 mm thick. Gamma quanta emitted by the target were recorded by two scintillation spectrometers with NaI(Tl) crystals, 70 mm in diameter and 100 mm thick; their resolution was 11–12 % for the ^{137}Cs line with energy $E_\gamma = 660\,\text{keV}$. The entire detector part of the system was separated

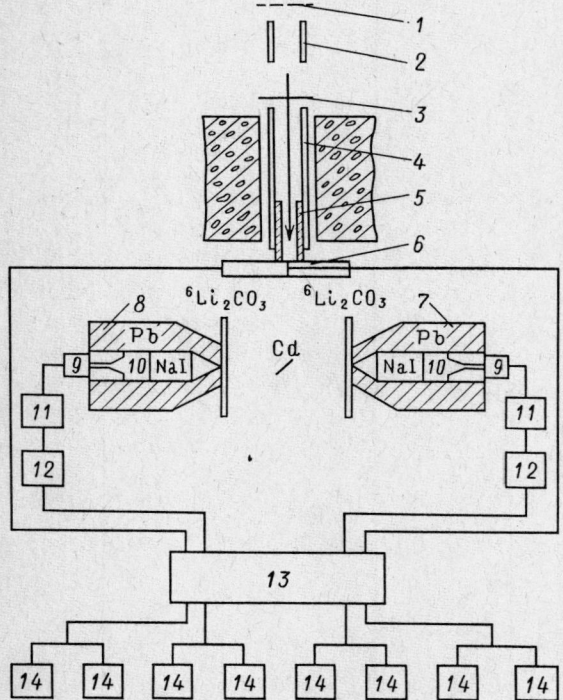

Fig. 8.2. The ITEP experiment analyzing asymmetry of gamma emission from polarized ^{114}Cd nuclei [146]: *(1)* depolarizing iron foil, *(2)* spin rotation magnet, *(3)* current-carrying foil, *(4)* magnetic guide, *(5)* collimator, *(6)* rotating iron foil depolarizer, *(7,8)* spectrometers, *(9)* cathode followers, *(10)* photomultipliers, *(11)* amplifiers, *(12)* analyzers, *(13)* commutator, *(14)* scalers

from the reactor hall by a thick concrete wall. Neutrons scattered by the target were absorbed by a layer of lithium carbonate enriched in ^{6}Li. Photomulitpliers and scintillation crystals were protected from magnetic fields by several steel and Permalloy shields and covered by a lead layer, 70 mm thick.

All the experiments singled out the same energy range of gamma-quantum pulses, corresponding to the 9.04 MeV transition in ^{114}Cd; namely, 8.5–9.5 MeV. Special attention was paid to eliminating the overlapping of gamma quanta which could bring gamma quanta of lower energies into the investigated energy range, and hence, reduce the effect under consideration.

The main difficulty in such experiments is the elimination of instrumental asymmetry due to the instability of either the electronics functioning or the neutron flux. The electronic equipment in the three experiments mentioned differed in circuit diagrams and in the implementation of several subunits.

Various methods were used to suppress instabilities. In the first experiment [146], fast comparison was carried out of the effects with the polarized and depolarized neutron beams. This was achieved by placing a revolving depolarizer (a circular disk with two opposing quadrants covered by iron foil and the other two left open) in the beam path (see Fig. 8.2). The neutron beam was interrupted by the foil, and hence, depolarized, 20 times a second [4]. The necessary energy interval was cut out of the pulse spectrum by differential discriminators. An electronic commutator fed the pulses successively into two scaler circuits corresponding to two states of polarization of the neutron beam. The neutron spins were flipped with respect to the direction of the permanent magnetic field on the target every 20 minutes, by a nonadiabatic spin flipper [18] (see Sect. 1.2.3).

In the second and third ITEP experiments [7, 19], fast comparison was carried out of the effects with the opposing directions of neutron spins. Correspondingly, a special rotation magnet and a nonadiabatic spin flipper [18] reversed the polarization in the beam at the target 10 times a second, for a fixed direction of the magnetic field. Measurements with the polarized and depolarized beams were alternated every 20 min. The depolarized beam experiment served to establish instrumental asymmetry. By combining the results obtained with the polarized and depolarized beams, it was possible to extract the sought value of the asymmetry coefficient.

A large number of control experiments were conducted in each run [154]. Thus, the asymmetry in the angular distribution of gamma quanta emitted by ^{114}Cd was measured in another energy range where the effect was expected to be greatly suppressed. Since in this interval, many transitions make their contributions, it was unlikely that the asymmetry would have an identical sign for all of them.

Control experiments were also run with other nuclei for which P-odd effects were not expected. Experiments with titanium and lead tested whether the apparatus was insensitive to the circular polarization of gamma quanta because the calculated circular polarization of the selected gamma quanta was equal to that of 9.04 MeV quanta emitted by ^{114}Cd nuclei.

A control experiment with a graphite target tested the sensititity of the system to neutrons scattered by the target. The asymmetry of the background was measured by removing the target from the beam. Another test was run with the vertical polarization of the neutron beam, in contrast to the main series of experiments in which neutron spins were oriented horizontally. In this experiment, $\cos\theta$ was zero, and the P-odd correlation was supposed to vanish. However, if the beam contains p-neutrons, the interference of the s- and p-levels can produce a P-even correlation of the type $s(p_n \times p_\gamma)$, where s is the neutron spin, and p_n and p_γ are the momenta of the neutron and gamma quantum, respectively (for details, see Chap. 10). In the ideal case, no such correlation occurs because the spin of the neutron is parallel (or antiparallel) to the gamma quantum momentum and $s(p_n \times p_\gamma) = 0$. The difference between the real geometry of the experiment and the ideal geometry could result in a correlation of this type. In control experiment, this correlation was enhanced by rotating the direction of polarization by 90°.

The results of all control experiments led to the conclusion that the asymmetry observed in the main run was caused by the angular asymmetry of the emission of 9.04 MeV gamma quanta after the capture of polarized thermal neutrons by ^{113}Cd nuclei.

The mean weighted value of the asymmetry coefficient a in three experiments was found to be $-(3.3\pm0.6)\times10^{-4}$. If this figure is corrected for 8.48 MeV gamma quanta falling within the selected energy interval (which reduce asymmetry, as we mentioned above) and for the overlapping pulses of lower-energy gamma quanta, the final result is [154]:

$$a = -(4.1\pm0.8)\times10^{-4} \ .$$

8.2.5 Other Experiments on the Analysis of P-Odd Effects in the ^{113}Cd$(n,\gamma)^{114}$Cd Reaction

The angular distribution of gamma quanta in the reaction ^{113}Cd$(\vec{n},\gamma)$ produced by polarized neutron beams was studied in a number of other laboratories. Systems at Ispra (Italy) [155] and Karlsruhe (FRG) [156] did not separate the $1^+ \to 0^+$ and $1^+ \to 2^+$ transitions; this is, as we have mentioned above, unacceptable. These experiments failed to find the asymmetry. The results obtained by the group in Risö (Denmark) [157, 158] do not contradict the ITEP results. They are given in Table 8.1 (see below).

Later, another group at ITEP [159] confirmed the results of the earlier experiments. These measurements served as test runs in the preparations for the experiment with ^{117}Sn nucleus.

The obtained value of the asymmetry coefficient of the angular distribution of gamma quanta in the reaction ^{113}Cd$(\vec{n},\gamma_0)^{114}$Cd, caused by polarized neutrons, is in good agreement with the results found by measuring the circular polarization of gamma quanta in the same reaction (reported by *Wilson* and his coworkers [160] who used a nonpolarized neutron beam). The expression for the circular polarization P_γ of the gamma quanta (emitted in nonpolarized neutron capture) due to the nonconservation of P parity (for a pure gamma transition),

$$P_\gamma = 2RF , \tag{8.10}$$

differs from the expression for the asymmetry coefficient (8.7) only in the absence of the spin factor A [149]. Consequently, the sign of the circular polarization is the same for all transitions originating from the same level of a nucleus (in addition, it is assumed that parities mix at the upper levels from which the electromagnetic transition takes place).

Therefore, the sign of the circular polarization of the gamma radiation emitted from ^{114}Cd nuclei is the same for the transitions with energies of 9.04 and 8.46 MeV. Wilson's group at Harvard University [160] made use of this fact by measuring the circular polarization of gamma rays with an energy of above 8 MeV, emitted in the $^{113}\text{Cd}(n,\gamma)^{114}\text{Cd}$ reaction with nonpolarized neutrons. The sign and magnitude of the circular polarization $P_\gamma = +(6.0\pm1.5)\cdot10^{-4}$ agree in sign and magnitude with the asymmetry coefficient a for 9.04 MeV gamma quanta emitted by ^{114}Cd nuclei, to within the experimental error (note that $A = -1$).

8.2.6 A Study of the P-Odd Effect in the Reaction $^{117}\text{Sn}(\vec{n},\gamma_0)^{118}\text{Sn}$

The asymmetry of gamma rays in the reaction $^{117}\text{Sn}(\vec{n},\gamma_0)^{118}\text{Sn}$ was studied both at ITEP and ILL. *Danilyan* et al. [159] used a beam of polarized thermal neutrons (see entry 1 in Table 1.1). The beam was aimed at a target of metallic tin enriched in ^{117}Sn up to 90 %. The target was monitored by four NaI(Tl) detectors placed on both sides of it, at the angles of 22.5 and 157.5° between the two detectors. The detector pulses were formed, energy-discriminated, and recorded by two groups of scalers (depending on the direction of neutron polarization). The horizontally directed polarization vector was stochastically flipped once a second (pulses from the neutron counter served as start pulses). Switching of the electronic recording channels after each exposure was also stochastic. Polarized- and nonpolarized beam measurements alternated every 16 min. The data supplied by scalers was analyzed by the on-line computer. A number of control experiments were carried out, among them, the measurement of the P-odd asymmetry in the reaction on ^{113}Cd, mentioned in Sect. 8.2.5.

Taking into account corrections for the neutron beam polarization, for the finite solid angle subtended by the gamma quantum detectors, and the measurement error with the depolarized beam, the authors of [159] arrived at the asymmetry coefficient $a = (8.1\pm1.3)\times10^{-4}$.

Wilson et al. [161] used a beam of polarized cold neutrons at the high-flux ILL reactor (entry 9 in Table 1.1). Two NaI(Tl) detectors were used for measuring the asymmetry coefficient. Neutron spins were flipped once a second. Control experiments were carried out with a ^{60}Co source, a depolarized beam, a neutron polarization vector rotation by 90°, and a change of range of gamma quantum energies. The asymmetry coefficient was found to be $a = (4.4 \pm 0.6) \times 10^{-4}$. The discrepancy of the values of a reported in [159] and [161] was not explained.

8.2.7 An Attempt to Study P-Odd Effects in Light Nuclei

The first attempt to study the asymmetry of gamma rays emitted by light nuclei was reported by *Wilson* and his coworkers [162], who used the reaction of polarized neutron capture by hydrogen.

The apparatus (Fig. 8.3) worked with the same polarized neutron beam as in the experiment with ^{117}Sn (entry 9 in Table 1.1). Neutron spins were reversed by one of the axial spin flippers [21] described in Sect. 1.2.3. The spins of neutrons leaving the spin flipper were then rotated through 90° in the plane perpendicular to neutron momenta.

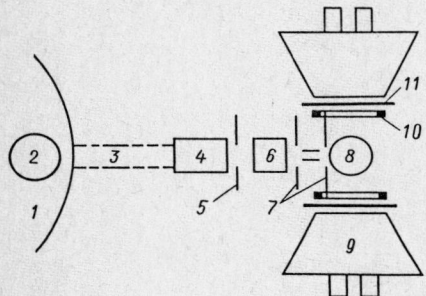

Fig. 8.3. The ILL experiment for measuring the asymmetry coefficient of gamma-ray emission in the radiative capture of polarized neutrons by protons [162]: *(1)* reactor, *(2)* liquid-deuterium moderator, *(3)* neutron guide, *(4)* polarizer, *(5)* lead shielding, *(6)* spin flipper, *(7)* collimators, *(8)* para-hydrogen target, *(9)* liquid scintillator, *(10)* coils producing constant magnetic field, *(11)* soft iron

A parahydrogen target (used to prevent depolarization in the scattering by orthohydrogen) was mounted in a cryostat with a volume of 23 liters. This target captured one half of the neutrons, and 2.23 MeV gamma quanta were emitted. Two liquid scintillation detectors, each with a volume of $0.5\,\mathrm{m}^3$, were used as gamma-ray detectors, and were monitored by four photomultipliers. The integral measurement technique was employed. The photomultiplier current was integrated over a period of 0.17 s, converted into a digital code several times within each run with the spin flipper (0.8 s), and then stored on magnetic tape. A large number of control experiments were carried out. The final result, i.e., the asymmetry coefficient $a = (0.06 \pm 0.21) \times 10^{-6}$, pointed to the absence of P parity nonconservation, within an error of 4×10^{-7}, in the reaction $\vec{n} + p \to d + \gamma$.

Recently, *Avenier* et al. [163] reported that Wilson's group at ILL succeeded in measuring the asymmetry of gamma ray emission in the reaction $\vec{n} + {}^2H \rightarrow {}^3H + \gamma$. The accuracy of the obtained asymmetry coefficient $a = (7.8 \pm 3.4) \times 10^{-6}$ is too low to speak about a discrepancy between the experiment and the theoretical prediction, $a \approx 10^{-6}$ [164].

8.2.8 Analysis of P-Odd Effects in the Integral Spectrum of Gamma Quanta

The anisotropy of gamma radiation emitted after polarized neutron capture was studied at LINP [165] in a number of nuclei. The difference from the work reported earlier was that no specific transition was singled out, but the effect was measured in the integral spectrum of gamma quanta emitted by a specified nucleus.

The apparatus for measuring the anisotropy and the measurement procedure are described in Sect. 12.2.2. A neutron beam ($\bar{\lambda} = 0.27$ nm), polarized perpendicularly to the neutron momentum, passed through a target protected by a layer of ^{6}LiF. Scintillation detectors of gamma rays were placed on both sides of the target. The results of measurements are shown in Table 8.1 in rows marked with "Integral spectrum".

8.2.9 Results of Studying the Anisotropy of Gamma Radiation Emitted by Nuclei After Polarized Neutron Capture

The results are summarized in Table 8.1. Note that the results for ^{114}Cd and ^{118}Sn are quite close to each other. This is not surprising because the mass numbers and decay schemes of the two nuclei are very similar.

Assuming the asymmetry found in the ^{114}Cd and ^{118}Sn nuclei to be caused by a P-odd interaction, and taking $R \approx 10^3$ for the enhancement factor, we obtain from (8.7) the amplitude of mixing for the states with opposite parities, $F \approx 2 \times 10^{-7}$, which agrees with the estimates of *Blin-Stoyle* [128], *Shapiro* [127], and *McKellar* [166]. The complexity of the nuclear states and the overlapping of their wave functions are such that only the order of magnitude of mixing amplitude can be theoretically estimated for the states with opposite parities.

So far, the experimentally obtained value of a in the reaction $\vec{n} + p \rightarrow d + \gamma$ is not sufficiently accurate for comparison with the theoretical prediction $a_{\text{theor}} \approx (0.5\text{--}0.6) \times 10^{-8}$ [124]. Nevertheless, the importance of this experiment can hardly be overestimated. *Danilov* [167] was able to show that the asymmetry of gamma ray emission as a result of polarized neutron capture by protons is sensitive to the isovector component ($\Delta T = 1$) of the potential of the nucleon-nucleon weak interaction, while the circular polarization of gamma quanta emitted in the capture of nonpolarized neutrons is sensitive to the isoscalar ($\Delta T = 0$) or isotensor ($\Delta T = 2$) component. Therefore, the experiments on measuring the asymmetry of gamma-ray emission and the circular polarization of gamma quanta, provided these experiments are sufficiently accurate, will make it possible to establish the isotopic structure of the nucleon-nucleon weak interaction.

Table 8.1. Asymmetry coefficients a of the gamma emission from nuclei after polarized neutron capture

Compound nucleus	Transition energy [MeV]	Characteristic of the regular transition	a, $[10^{-6}]$	Source
^{2}H	2.23	$0^+ \xrightarrow{M1} 1^+$	0.06 ± 0.21	ILL [162]
^{3}H	6.24	$\frac{3}{2}, \frac{1}{2}^+ \xrightarrow{M1} \frac{1}{2}^+$	7.8 ± 3.4	ILL [163]
^{36}Cl	Integral spectrum	–	-27.8 ± 4.9	LINP [165]
^{80,82}Br	Integral spectrum	–	-19.5 ± 1.6	LINP [165]
^{114}Cd	9.04	$1^+ \xrightarrow{M1} 0^+$	-410 ± 80	ITEP [154]
			-250 ± 220	Risö [157]
			-60 ± 180	Risö [158]
			-500 ± 120	ITEP [159]
^{118}Sn	9.31	$1^+ \xrightarrow{M1} 0^+$	810 ± 130	ITEP [159]
			440 ± 60	ILL [161]
	Integral spectrum	–	2.4 ± 1.6	LINP [165]
^{140}La	Integral spectrum	–	-17.8 ± 2.2	LINP [165]

The LINP experiment on measuring the circular polarization of gamma quanta in the $np \to d\gamma$ reaction on nonpolarized neutrons gave the upper limit $P_\gamma \leq 5 \cdot 10^{-7}$ [168].

The interpretation of the effects found in the integral spectrum of gamma quanta [165] meets with obstacles because of the paucity of data on the characteristics of γ-transitions in the nuclei investigated in (n, γ)-reactions. In principle, the effects must be of opposite signs for the γ-transitions from the compound nucleus to states with different spins, and thus must substantially cancel one another out; i.e., the net effect in the integral spectrum is expected to be greatly reduced. The experiment demonstrates that this is not the case. Consequently, the existing notions on the mechanisms of enhancement of P-parity nonconservation effects in (n, γ)-reactions and on the mechanisms of mixing of highly excited states need further elaboration.

Note added in proof: The asymmetry coefficient $a = (157 \pm 53) \times 10^{-6}$ has been measured at ILL [M. Avenier, G. Badieu, H. Benkoula, J.F. Cavaignac, A. Idrissi, D.H. Koang, R. Wilson: Nucl. Phys. **A 436**, 83 (1985)] for the most energetic transition in the reaction ^{35}Cl$(\vec{n}, \gamma_0)^{36}$Cl with the energy of 8.58 MeV. The asymmetry coefficient $a = -(21.2 \pm 1.7) \times 10^{-6}$ has also been measured in the same reaction but in the integral spectrum of gamma quanta.

8.3 Experimental Investigation of Time Reversal Invariance in Nuclear Interactions

8.3.1 Introductory Remarks

Immediately after the discovery of CP violation in K^0 meson decays [65], a number of various hypotheses were advanced as to which interaction is responsible for this violation. Thus, if it were found that CP parity is broken by the hadron-hadron or hadron-photon interaction conserving strangeness and spatial parity, then T-noninvariant effects in electromagnetic decays of nuclei at a 0.1% level could be expected on the basis of the CPT theorem [169] (see Sects. 6.1 and 6.2).

At present, the violation of CP parity in K^0 meson decays is explained by a mechanism of superweak mixing of K_2^0 and K_1^0 mesons with a coupling constant of about 10^{-9} G, where G is the universal Fermi constant [58]; in this case, no T-noninvariant effect can be detected in any currently feasible nuclear-physics experiments. In fact, this explanation is a result of analysis of experimental data, including those obtained in polarized-neutron experiments to be described below.

8.3.2 Fundamentals of the Method

In order to study the invariance under time reversal, various T-noninvariant angular correlations in cascade nuclear gamma transitions are analyzed [170] (see also [148]).

The general correlation function for a polarized nucleus emitting two cascade gamma quanta is given in [148, 170]. If an interaction is invariant under time reversal, the phases of the reduced nuclear matrix elements describing the angular correlations of the consecutive gamma transitions can be chosen equal, and the matrix elements themselves can be assumed to be real [171]. If T-invariance is broken, a nonzero (or distinct from π) phase difference η between matrix elements appears; this can be detected by measuring the interference terms in the angular correlation of mixed gamma transitions. The mixing parameter, that is, the ratio of the reduced matrix elements of transitions between nuclear states with spins J_0 and J_1,

$$\delta_1 = \langle J_1 | L+1 | J_0 \rangle / \langle J_1 | L | J_0 \rangle = |\delta_1| \exp i\eta \tag{8.11}$$

is then complex (L and $L+1$ stand for the multipolarity of the radiation).

A mixed transition in the cascade is a necessary condition for the appearance of the interference term containing the phase difference η. No interference term arises for a cascade of pure transitions because the emission intensity is given by squared moduli of the matrix elements of the transitions. Experiments have maximum sensitivity to the interference of unequal-multipolarity radiations if the two emissions have nearly equal amplitudes, i.e., if $|\delta_1| \approx 1$. Therefore, the phase difference η characterizes the strength of the interaction which violates the T invariance.

It was suggested [172] that polarized neutrons can be used for studying the T invariance of nuclear forces; namely, that the T-noninvariant (P-even) correlation (see Table 6.1)

$$(\boldsymbol{p}_1\boldsymbol{p}_2)\boldsymbol{J}(\boldsymbol{p}_1\times\boldsymbol{p}_2)\propto \sin 2\theta ,\qquad(8.12)$$

arising in a mixed gamma transition, be measured, and that the nuclei with spin J in the initial state be polarized by irradiating them with polarized neutrons ($\boldsymbol{p}_1$ and $\boldsymbol{p}_2$ are the momenta of the first and second gamma quanta of the cascade, and θ is the angle between the vectors $\boldsymbol{p}_1$ and $\boldsymbol{p}_2$).

The angular distribution of two gamma quanta of the cascade, successively emitted by a nucleus after polarized neutron capture, can be written in the form [173]:

$$w(\theta) = \text{const}\,[1 + A_2 P_2(\cos\theta) + A_4 P_4(\cos\theta) \\ + BP_n\delta_1 \sin\eta \sin 2\theta \cos\varphi]\;.\qquad(8.13)$$

Here, A_k denotes the coefficients of the angular distribution; $P_k(\cos\theta)$ are the Legendre polynomials; θ is the same angle as in (8.12); B is the coefficient in front of the T-nonvariant term, depending on the spins of the levels, on the mixing parameters of the two transitions (if the first gamma transition is nonmixed or the spin of the intermediate state J_1 equals either 0 or 1/2, then $B = 0$), and on the geometry of the experimental setup (solid angles between the target and the detector); P_n is the polarization of the neutron beam; and φ is the angle between the normal to the plane of gamma-ray emission and the direction of neutron beam polarization.

As follows from (8.13), time reversal is equivalent to the replacement $\theta \to -\theta$, which is readily achieved by reversing the beam polarization at a fixed angle θ. The last term in (8.13) then reverses its sign. The principle of the experiment is, therefore, clear. The neutron polarization in the beam incident on the target must be perpendicular to the plane of gamma-ray emission ($\varphi = 0$). Two gamma-quantum detectors are installed in the plane perpendicular to polarization direction, at a $\theta = 45$ or $135°$ angle to each other. The T-noninvariant term is thereby maximized. Coincidence numbers for gamma quanta of two transitions are recorded: $J_0\to J_1$ and $J_1\to J_2$ for two opposite directions of the neutron beam polarization. This experiment yields the asymmetry coefficient of gamma-ray emission,

$$a = \frac{w(\theta) - w(-\theta)}{(1/2)[w(\theta) + w(-\theta)]} = P_n E \sin\eta ,\qquad(8.14)$$

from which the magnitude and the sign of $\sin\eta$ can be found. The following notation was introduced in (8.14):

$$E = \frac{2B\delta_1 \sin\theta \cos\theta \cos\varphi}{1 + A_2 P_2(\cos\theta) + A_4 P_4(\cos\theta)}\;.\qquad(8.15)$$

The possibility of the imitation of T-noninvariance by the effects of exchange of virtual gamma quanta between the initial, intermediate, and final states calls for special attention. Computations show that in most gamma transitions of interest, the contribution to asymmetry by the virtual gamma-quantum exchange is very small in comparison with currently achieved experimental errors [174].

8.3.3 Experiments

A typical difficulty for experimental work in this field is the small number of nuclei for which the mixed emission after neutron capture was established. Furthermore, experiments must meet additional requirements: the mixing parameter must be close to unity; a minimum (and known) number of transitions from the upper excited level of the compound nucleus must occur prior to the mixed transition, in order for the polarization of the initial state with spin J_0 not to be close to zero; moreover, the cascade to be analyzed must have an appreciable intensity and be reliably energy-resolved by the detectors.

The same paper [172] suggested that the ^{49}Ti nucleus be used. The starting level of the cascade is then the 1.723 MeV level. The 1.723 MeV $\to$ 1.381 MeV transition is a mixed $M1 + E2$ transition with the mixing parameter $\delta_1 = -0.1$. The mixing parameter of this transition was measured at ITEP in order to study the T-noninvariant correlation [175]. The mixing parameter being low, the coefficient E is small: $E = 0.016$. It is not surprising, therefore, that the accuracy of the experimental value of $\sin \eta$ reported for this nucleus at the Nuclear Research Institute (NRI) in Řež (Czechoslovakia) [176] proved to be insufficient: $\sin \eta = (17\pm25)\times10^{-3}$.

The same cascade was studied at ILL on a modernized apparatus containing six NaI(Tl) detectors and detecting eight possible coincidence combinations [177]. The result (see Table 8.2) is still insufficiently accurate, despite the error reduced by a factor of 3.

The ^{36}Cl nucleus proved to be more suitable. A simplified transition scheme of this nucleus is shown in Fig. 8.4. For the 8.583 MeV $\to$ 0.788 MeV $\to$ 0 MeV cascade of this nucleus, $E = 0.162$. Therefore, for this cascade, the coefficient in front of $\sin \eta$ is greater by an order of magnitude than for the cascade of ^{49}Ti nuclei. The error in the measured value of $\sin \eta$ could then be reduced.

The first experiment on measuring the T-noninvariant angular correlation of gamma quanta emitted by ^{36}Cl nuclei was carried out at Karlsruhe (FRG) [178]. Similar measurements with ^{36}Cl, but on a higher-aperture system were carried out at ITEP [179] (Fig. 8.5).

The apparatus used one of the ITEP polarized neutron beams (entry 1 in Table 1.1). The target was a thin aluminum container with C_2Cl_6 powder. The neutron radiative capture γ-quanta emitted by ^{35}Cl were recorded by four NaI(Tl) detectors, 70 mm in diameter and 100 mm thick, placed in the plane perpendicular to the beam polarization, so that the average angle between two detectors was 45 or 135°. The photomultipliers were thoroughly protected from stray magnetic fields by shields of soft steel and Permalloy.

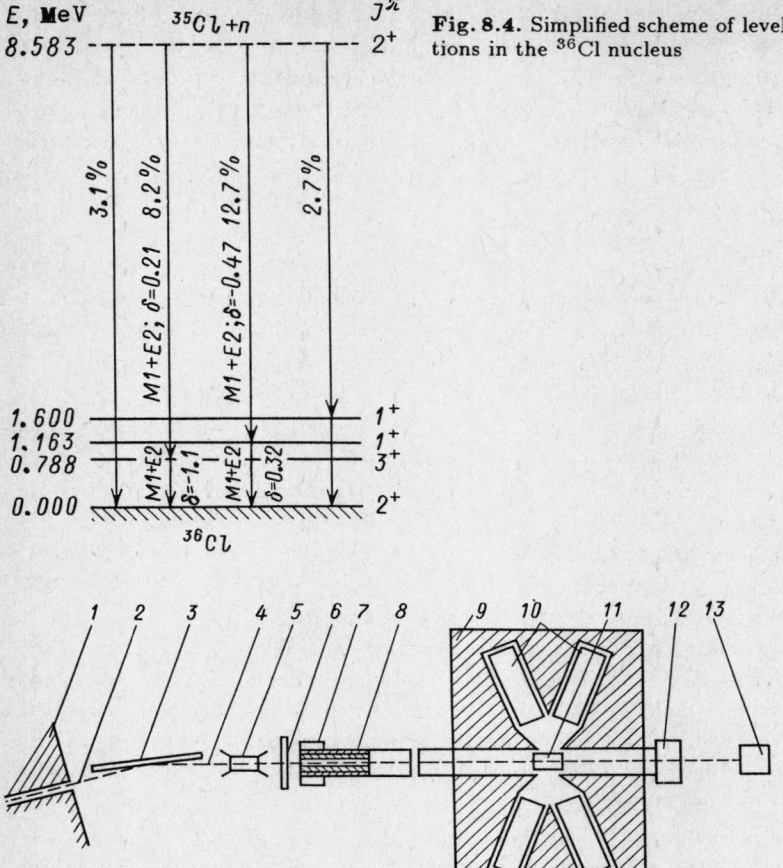

Fig. 8.4. Simplified scheme of levels and transitions in the ^{36}Cl nucleus

Fig. 8.5. The ITEP experiment measuring T-nonvariant correlations [179]: *(1)* reactor, *(2)* neutron beam, *(3)* cobalt mirror, *(4)* polarized neutron beam, *(5)* spin rotation magnet, *(6)* nonadiabatic spin flipper, *(7)* leading field magnet, *(8)* collimator, *(9)* shielding, *(10)* gamma-ray detectors, *(11)* target, *(12)* analyzer magnet, *(13)* neutron counter

The neutron beam polarization was flipped once a second because this operation simultaneously averaged out the possible fluctuations of instrument parameters. Each of the gamma quantum detectors recorded both the low- and the high-energy components of the cascade, 0.79 and 7.79 MeV. Hence, four detectors provided eight coincidence combinations; i.e., eight values of the asymmetry coefficient were measured simultaneously, by taking into account the flipping of the neutron beam polarization vector. These eight coincidence combinations were recorded by multichannel electronic equipment. The resolving time τ of the fast coincidence circuit was 15 ns. The fact that each detector participated in two pairs of coincidence combinations, giving opposite-sign contributions to asymmetry, reduced the requirements for electronic stability.

This was essential in view of a rather high loading of photomultipliers (2×10^5 pulses/s).

The mean weighted asymmetry coefficient was $a = (-0.9\pm1.1)\times10^{-4}$. Measurements demonstrated that the coincidence peak of the 8.58 MeV → 0.79 MeV → 0 MeV cascade, averaged over eight combinations, is composed of a 38 % contribution by 0.79 MeV gamma quanta, a 28 % contribution by a closest cascade 8.58 MeV → 1.16 MeV → 0 MeV, and a 34 % contribution by spurious coincidences background.

The mean weighted value of E for these three components was found to be equal to 0.051. Correspondingly, the phase difference for the indicated matrix elements in the 7.79 MeV mixed electromagnetic transition in ^{36}Cl is found to be $\eta \approx \sin\eta = (-1.8\pm2.2)\times10^{-3}$.

8.3.4 Measurement Results for Angular Correlations of Gamma Quanta Emitted by Nuclei After Polarized Neutron Capture

The summary of results shown in Table 8.2 leads to the conclusion that the upper bound on the phase difference of the matrix elements of the mixed gamma transitions is $-(2\pm2)\times10^{-3}$. This value must be compared with the results of the highest-precision measurements of the same phase difference in experiments of another type, also employing mixed gamma transitions.

The highest precision in measuring the γ-γ-correlation was achieved with the ^{110}Cd nucleus (the $3^+ \to 2^+$ transition at 1505 keV and $\delta = -1.09$), polarized by the low-temperature method. The result obtained was $\sin\eta = (1.5\pm2.2)\times10^{-3}$ [180]. The highest-precision measurements of the T-noninvariant correlation using the Mössbauer effect in the ^{99}Ru nucleus (the $3/2^+ \to 5/2^+$ transition at 90 keV and $\delta = -1.64$) yielded $\sin\eta = (1.0\pm1.7)\times10^{-3}$ [181].

Cheung et al. [182] achieved still higher precision in measuring the angular distribution of linearly polarized gamma quanta emitted from nuclei polarized by the low-temperature method. Nevertheless, care must be exercised in interpreting the results of studies on low-energy gamma quanta, in view of the data published in [183], where it was demonstrated that phase differences on the order of 10^{-3} may arise, owing to virtual processes of internal conversion of low-energy gamma quanta on atomic shell electrons.

Table 8.2. The values of the sine of phase difference between matrix elements of two successive gamma transitions

Compound nucleus	Transition characteristics	$\sin\eta$, $[10^{-3}]$	Source
^{49}Ti	$1/2^- \xrightarrow[E=0.34\text{ MeV}]{\delta_1=-0.1} 3/2^-$	17 ± 25	NRI, Rez [176]
		-9 ± 7	ILL [177]
^{36}Cl	$2^+ \xrightarrow[E=7.79\text{ MeV}]{\delta_1=0.21} 3^+$	4 ± 12	Karlsruhe [178]
		-1.8 ± 2.2	ITEP [179]

Another type of experiment consists in measuring the angular γ-γ-correlations for polarized nuclei in which one of the transitions is greatly suppressed. Thus, a strongly suppressed 922 keV transition in ^{184}W nuclei was investigated [184]. However, the causes of this strong suppression of gamma transitions are not sufficiently clear, so that the conclusions drawn from experiments with strongly suppressed gamma transitions do not have solid theoretical backing.

To recapitulate, the measurements of T-noninvariant correlations in gamma rays emitted by nuclei yield an upper bound on the T-noninvariant component of the interaction potential: $F' \leq 1 \times 10^{-3}$. These data and the conclusions on the upper bound of 6×10^{-25} e×cm on the electric dipole moment of the neutron [80] (see Sect. 2.3.4) cast substantial doubt on the hypothesis of the violation of CP invariance in hadron-hadron and hadron-photon interactions. Conversely, the hypothesis of the violation of CP invariance in the superweak interaction appears to gain additional support.

9. Anisotropy of Beta Particles Emitted by Nuclei After Polarized Neutron Capture

Anisotropy is a characteristic feature of the angular distribution of beta particles emitted after polarized neutron capture in reactions involving polarized neutrons. This phenomenon can be used to measure the spin characteristics of compound nuclei. If the method is supplemented with the resonance depolarization of the created beta-radioactive nuclei, it is possible to determine the magnetic moments of some short-lived nuclei formed as a result of neutron capture. Studying the anisotropy of beta emission proved to be especially fruitful for a number of problems in solid state physics. Here we deal only with the nuclear-physics applications of this phenomenon.

9.1 Fundamentals of the Experimental Method

9.1.1 Polarization of Nuclei

As a result of polarized neturon capture by nonpolarized nuclei with spins J_i, partially polarized compound nuclei with spins $J_c = J_i \pm 1/2$ are formed. Compound nuclei remain largely polarized after they emit gamma quanta. The degree of polarization of the ground state with spin J_f is dictated by the spins of the intermediate states J', J'', and the multipoles of the gamma transitions. If these compound nuclei are beta-radioactive, their beta emission has angular anisotropy by virtue of the nonconservation of spatial parity under the weak interaction.

The process involved here (Fig. 9.1) is an extension of the one used to analyze the anisotropy of the gamma radiation from polarized nuclei after polarized neutron capture (see Fig. 4.1).

In order to derive the expression for the angular anisotropy of beta emission from a compound nucleus in its ground state, we need to know its polarization. When nonpolarized nuclei capture completely polarized s-neutrons ($P_n = 1$), the polarization of compound nuclei in the excited state is [185]:

$$P_N = \begin{cases} P_+(J_c) = (1/3)[1 + 2/(2J_i + 1)] & \text{if } J_c = J_i + 1/2 \ , \\ P_-(J_c) = -1/3 & \text{if } J_c = J_i - 1/2 \neq 0 \ , \\ P_-(J_c) = 0 & \text{if } J_c = 0 \ . \end{cases} \quad (9.1)$$

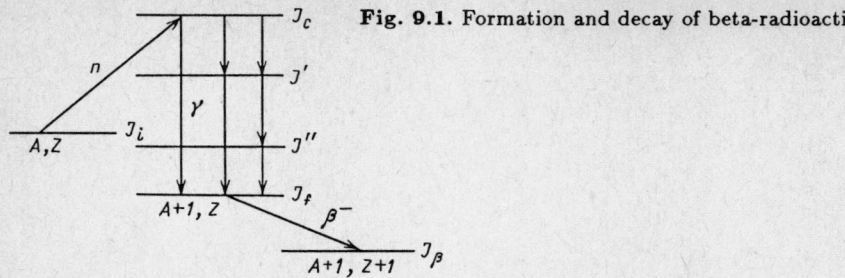

Fig. 9.1. Formation and decay of beta-radioactive nuclei

In a number of nuclei, thermal neutrons are captured into both allowed spin channels. Denoting by w_+ and w_- the probabilities of capture into the channels $J_i + 1/2$ and $J_i - 1/2$, respectively, we write the expression for the polarization of nuclei in the final state, $P_0(J_f)$, as a function of the polarization of compound nuclei in the excited state [186] in the form:

$$P_0(J_f) = w_+ \bar{a}_+ P_+(J_c) + w_- \bar{a}_- P_-(J_c) ,$$
$$w_+ + w_- = 1 . \tag{9.2}$$

The coefficients $\bar{a}_+$ and $\bar{a}_-$ determine the decrease in nuclear polarization due to gamma emission, and are given by the expression

$$\bar{a}_\pm = \sum_i \left[\omega_i \left(\prod_k a_k \right)_i \right]_\pm , \tag{9.3}$$

where the summation is taken over the cascades, ω_i is the fraction of intensity for the ith cascade, $\prod_k a_k$ are the products over the gamma transitions in the cascade, and the coefficients a_k for each individual gamma transition depend on the multipoles of gamma rays and on the spins of the levels between which the gamma transition occurs.

The general expressions for the change in polarization due to emission are given in [151]. *Shapiro* [185] derived the formulas giving the values of a_k for dipole and quadrupole transitions (if the gamma transition is mixed, the value of the coefficient a_k is a weighted mean of the two modes of emission). The numerical values of a_k for a number of transitions are given in Table 9.1 [186]. Obviously, the polarization in the final state is zero if one of the spins J_c or J_f is zero.

Table 9.1. Values of the coefficients a_k

Multipolarity of the radiation	$J' \to J''$ gamma transition			
	$1 \to 2$	$2 \to 1$	$1 \to 1$	$2 \to 2$
$E1$ or $M1$	3/4	1	1/2	5/6
$E2$	1/4	1/3	$-1/2$	1/2

If the gamma transition schemes and level spins are known, it is possible to calculate the probabilities w_+ and w_-, i.e., to find the contributions of the two possible spin channels to the cross section of thermal neutron capture by a given nucleus. This is done by the formulas (9.2) after calculating the coefficients $\bar{a}_+$ and $\bar{a}_-$ and experimentally determining the polarization of beta-radioactive nuclei, $P_0(J_f)$ (see below).

9.1.2 Angular Anisotropy of the Beta Emission

The angular distribution of beta particles emitted by polarized nuclei after polarized neutron capture (beam polarization P_n) is given by the expression

$$w(\theta) = \text{const}(1 + a\cos\theta) , \qquad (9.4)$$

where $a = (v/c)P_n P_0(J_f)\beta$, θ is the angle between the neutron polarization vector and the momentum of the beta particle, v/c is the relative velocity of the beta particle, $P_0(J_f)$ is the final state polarization of beta-radioactive nuclei, and β is a coefficient which depends on the spin J_β of the nuclei formed after the decay and on the type of the beta transition. The expression (9.4) is identical to (6.1), which describes the asymmetry of electron emission in the beta decay of polarized nuclei. This similarity is natural because both phenomena stem from the violation of spatial parity in the weak interaction (see Sect. 6.2).

Let us tabulate the values of β as a function of spin J_β for allowed beta transitions [185, 186]:

J_β	β
$J_f - 1$	± 1
J_f	$\pm \dfrac{1}{J_f + 1} \dfrac{1 + 2\sqrt{J_f(J_f+1)}X}{1 + X^2}$
$J_f + 1$	$\mp \dfrac{J_f}{J_f + 1}$

Here, $X = C_V M_F / C_A M_{GT}$, where C_V and C_A are the constants of the vector and axial-vector variants of the weak interaction, and M_F and M_{GT} are the nuclear matrix elements of the vector (Fermi) and axial-vector (Gamov-Teller) variants of the theory (see Sect. 6.4). The plus (minus) sign corresponds to the emission of positrons (electrons).

9.1.3 Experimentally Measured Asymmetry of the Beta Emission

The experimental value of the asymmetry coefficient of the beta emission is smaller than the theortical prediction, for the following three reasons [186]:

1. electron backscattering on the target and on the objects surrounding it;
2. finite angular size of the target and of the detector of beta particles;

3. various relaxation processes that reduce the nuclear polarization $P_0(J_f)$ with time.

The first factor is taken into account by introducing a correction for backscattering F, and the second, by introducing the effective mean cosine $\Omega = \overline{\cos\theta}$, which characterizes the reduction in asymmetry as a result of (a) electrons being emitted at various angles θ and (b) finite thickness of the target.

It was shown [186] that in a sufficiently thick specimen,

$$\Omega = 2(1 - \cos^3\theta_1)/3(1 - \cos^2\theta_1) , \qquad (9.5)$$

where θ_1 is one-half of the angle subtended by the detector, as seen from a pointlike target. For an infinitely thin specimen,

$$\Omega = (1 + \cos\theta_1)/2 . \qquad (9.6)$$

The difference between the values of Ω yielded by (9.5) and (9.6) is negligible up to $\theta_1 = 50°$. In this case, the number of decay electrons recorded in a given solid angle along the direction of neutron polarization and against this direction is

$$N_{0,\pi} = N_0(1 \pm \varepsilon_0) , \qquad (9.7)$$

where the asymmetry ε_0 is

$$\varepsilon_0 = \overline{(v/c)} P_n P_0(J_f) \beta \Omega F . \qquad (9.8)$$

The bar over v/c in (9.8) denotes averaging of the velocity of the electrons over the spectrum, taking into account the self-absorption in the target.

The third factor contributing to the reduction of asymmetry is accounted for by introducing a correction for polarization relaxation:

$$P(J_f) = P_0(J_f) \exp(-t/T_1) , \qquad (9.9)$$

where $P_0(J_f)$ is the polarization of beta-radioactive nuclei at the moment of formation, $P(J_f)$ is the value of this polarization at a time t, and T_1 is the relaxation time of the system of polarized beta-radioactive nuclei.

A system of polarized beta-radioactive nuclei is produced in experiments over a finite interval of time t_0 during which the target is irradiated by polarized neutrons. Then the asymmetry of the beta decay is measured during a time Δt (beginning at an instant t). If we take into account the beta decay of nuclei at a time constant λ, then the experimentally observed asymmetry $\varepsilon(t_0, t, \Delta t)$ is related to its initial value ε_0 by the formula [186]:

$$\varepsilon(t_0, t, \Delta t) = \varepsilon_0 \frac{\lambda^2}{[\lambda + (1/T_1)]^2} \frac{1 - \exp\{-[\lambda + (1/T_1)]t_0\}}{1 - \exp(-\lambda t_0)}$$
$$\times \frac{1 - \exp\{-[\lambda + (1/T_1)]\Delta t\}}{1 - \exp(-\lambda \Delta t)} \exp(-t/T_1) . \qquad (9.10)$$

The relaxation time T_1 and the value of ε_0 are found by measuring ε as a function of time and making use of (9.10); then $P_0(J_f)$ is found from (9.8).

9.1.4 Nuclear Magnetic Resonance of Polarized Beta-Radioactive Nuclei and the Measurement of Nuclear Magnetic Moments

The nuclear g-factors, and hence, the nuclear magnetic moments, can be measured by depolarizing the partially polarized beta-radioactive nuclei (produced as a result of polarized neutron capture) by the magnetic resonance depolarization method (as described in Sect. 2.1 for free polarized neutrons), and then monitoring the decay of beta-emission anisotropy [185]. This is achieved by applying an oscillating rf field at a frequency varied in a wide range perpendicularly to a static uniform field H_0, which is parallel to the neutron polarization vector. At a frequency equal to the Larmor value $\omega_L = \gamma H_0$ ($\gamma = \mu/J_f$ is the gyromagnetic ratio for the beta-radioactive nucleus, and μ and J_f are the magnetic moment and spin, respectively, of this nucleus), quantum resonance transitions are induced between two nuclear energy levels in the field H_0. The magnitude of polarization of the beta-radioactive nuclei decreases.

It was shown [187] that asymmetry ε is given as a function of field frequency ω by a formula similar to (9.10), with $1/T_1$ replaced by $(1/T_1 + 2w)$. The quantity $1/2w$ is the relaxation time for the system of polarized beta-radioactive nuclei placed in an rf field. If the spin J_f of a beta-radioactive nucleus is known, its magnetic moment is given by

$$\mu = \omega_L J_f / H_0 . \tag{9.11}$$

This method makes it possible to measure the magnetic moments of short-lived nuclei with a half-life of up to several minutes; no other technique is effective in this range. Note that nuclei with $J_c = 0$, which are of no interest in the study of the spin chracteristics of compound nuclei, can be used to measure the magnetic moments.

9.1.5 Measurements of Quadrupole Moments of Nuclei

Quadrupole moments of nuclei can be estimated from the experimental data on the relaxation time T_1 of a system of polarized beta-radioactive nuclei measured as a function of the target temperature T. The relaxation rate can be written, under certain assumption, in the form [188]:

$$1/T_1 \sim \frac{2J_f + 3}{J_f^2(2J_f - 1)}(\gamma Q)^2 F(T/\Theta) , \tag{9.12}$$

where γ is the proportionality coefficient, Q is the quadrupole moment of the relevant nucleus, and Θ is the Debye temperature of the target. This formula is based on the general theory of quadrupole relaxation [189] which predicts the following relation for two isotopes, A and B, of the same element in the same chemical compound at a given temperature:

$$R(A,B) = \frac{(T_1)_A}{(T_1)_B} = (Q_B/Q_A)^2 \frac{f(J_{fB})}{f(J_{fA})}, \qquad (9.13)$$

where $f(J_{fi}) = (2J_{fi}+3)/J_{fi}^2(2J_{fi}-1)$.

The quadrupole moment of the isotope B can be found from (9.13) if that of the isotope A is known and the relaxation times $(T_1)_A$ and $(T_1)_B$ of the two isotopes in the same compound have been measured.

Another method of determining the quadrupole moments of short-lived nuclei consists in measuring the quadrupole splitting of nuclear magnetic resonance spectra. We shall not describe this technique, but refer the reader to its description in [190].

9.2 Survey of Experiments

The first experiments on measuring the beta emission anisotropy were carried out in 1957 [191] immediately after the discovery of parity nonconservation in the weak interaction. The authors of [191] studied the capture of polarized neutrons by a Li_2CO_3 target and the decay of ^8Li nuclei. *Connor* [192] employed the resonance depolarization method and measured the gyromagnetic ratio for ^8Li, and hence, the magnetic moment of ^8Li. The results of these measurements are summarized in Table 9.2. *Wapstra* and *Connor* [193] studied the decay of the ^8Li, ^{20}F, and ^{28}Al nuclei, and demonstrated the importance of placing the target in a permanent magnetic field of about 240 A/cm which prevents nuclei from depolarization. Moreover, they concluded that in order to reduce the gamma quanta background, the target must be intermittently irradiated.

Table 9.2. Values of the probability of thermal neutron capture, w_+, into the $J_c = J_i + 1/2$ channel, the magnetic moments of nuclei, μ, and quadrupole moments of nuclei, Q,

Compound nucleus	w_+, [%] [186]	μ/μ_N	Q, $[10^{-24}$ cm$^2]$
^8Li	>86	1.6530(8) [186, 192]	$Q(^8Li)/Q(^7Li)=$ 0.78(1) [190]
^{20}F	>42	2.0925(4) [186, 193]	0.064(12) [200]
^{28}Al	–	2.789(1) [199]	–
^{38}Cl	–	2.05(2) [199]	–
^{66}Cu	>9	–	–
^{108}Ag	>24	2.6727(10) [201]	–
^{110}Ag	100	2.7111(10) [201]	$Q(^{110}Ag)/Q(^{108}Ag) = 1.49(7)$ [201]
^{116}In	–	2.7723(10) [188, 198] 2.7701(8) [199]	0.09(2) [188]

Note. The numbers in parentheses indicate uncertainty regarding the last significant digits of the reported result

Abov et al. [194] applied this technique to compound nuclei ^{8}Li, ^{108}Ag, and ^{110}Ag. They were the first to combine target cooling to down the temperature of liquid helium with a magnetic field to eliminate the depolarization of nuclei.

Connor and *Tsang* [195] replaced the oscillating rf field in the apparatus described in [193] by the rotating field of two rf coils, by analogy to the work on determining the sign of the magnetic moment of the neutron (see Sect. 2.2.4). They showed [195] that the magnetic moment of ^{8}Li is positive, and measured the magnitude and sign of the magnetic moment of the ^{20}F nucleus.

Let us look at the details of the work (Fig. 9.2) carried out at ITEP by *Gulko* et al. [186]. The ITEP polarized neutron beam (entry 1 of Table 1.1) was used. A metal cryostat was employed to cool the targets to the temperature of liquid helium. The specimen to be studied was glued to the magnesium cooling guide which was bolted to the cryostat. The neutron beam was sent onto the specimen, and decay electrons reached the detectors through windows made of thin copper foil. In order to prevent the collapse of the copper foil under the external atmospheric pressure, the lower part of the cryostat was placed in an evacuated chamber. The size of the specimen was 60×60×4 mm.

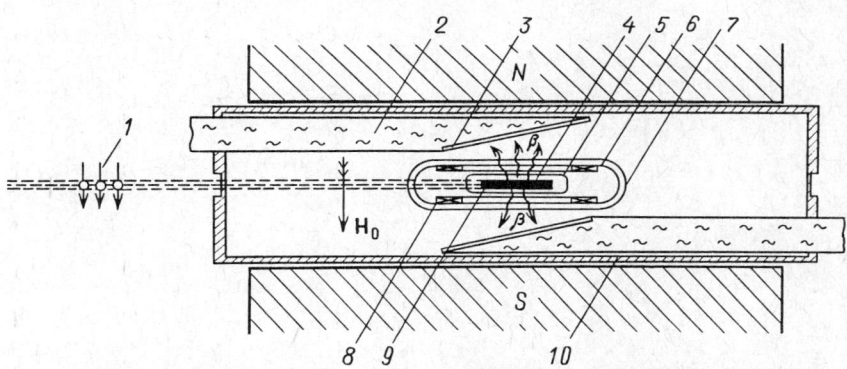

Fig. 9.2. The ITEP apparatus for studying the anisotropy of beta emission from nuclei after polarized neutron capture [186]: *(1)* polarized neutron beam, *(2)* light guide, *(3)* scintillator, *(4)* target, *(5)* rf coil, *(6)* nitrogen shield of cryostat, *(7)* thin-wall shield, *(8)* coil for field modulation, *(9)* magnesium cooling guide bar in contact with liquid helium reservoir, *(10)* vacuum chamber

The cryostat with the chamber was mounted in the pole gap of the magnet, producing a static magnetic field of up to 4 kA/cm, with nonuniformity not exceeding 10^{-4}. The rf magnetic field destroying the polarization of beta-radioactive nuclei was produced by a coil surrounding the specimen. The frequency range of the coil was 0.15–20 MHz, with field strength amplitude of 1–2.4 A/cm. The nuclear magnetic resonance line was broadened through a modulation of static magnetic field by passing ac current through two Helmholtz coils placed on both sides of the specimen.

The decay electrons emitted from the specimen were detected by two plastic scintillators, 70 mm in diameter an 1 mm thick. Scintillator flashes were transferred to photomultiplier cathodes by Plexiglas light guides, 450 mm in length each.

The asymmetry was calculated using the formula

$$\varepsilon = \frac{\sqrt{R}-1}{\sqrt{R}+1}, \quad R = \frac{N_1'/N_2'}{N_1/N_2}, \qquad (9.14)$$

where N_1 and N_2 are the counting rates with neutron spins directed from the first to the second counter. For opposing orientation of spins, the rates are denoted by N_1' and N_2'. The specimen was irradiated by neutrons having a specified direction of polarization. The beam was then rapidly cut by a chopper, and the two decay electron counters were switched on. The counting circuit made it possible to carry out the measurement during a prescribed time interval Δt at any moment during the whole decay time of the nuclei analyzed. After the decay of beta-radioactive nuclei was completed, the cycle was repeated with the opposite direction of neutron polarization.

In the measurements of the beta-decay asymmetry as a function of frequency of the rf field, the static magnetic field was maintained constant, and the specimen was subjected to the rf field continuously over the whole period of irradiation and measurement.

The relative electron velocity v/c was calculated from the known beta spectrum, taking into account its distortion due to self-absorption of electrons in the specimen. The function Ω was found using the formula (9.5). As a result of focusing of the electrons in the static magnetic field, the angle θ_1 was a function of H_0, changing from 55° at $H_0 = 0$ to 90° at large values of H_0. In a given field H_0, the angle θ_1 was found from the experimental dependence of the counting rate on the field, $N(\theta_1)/N(90°) = 1 - \cos^2\theta_1$, the specimen being irradiated by nonpolarized neutrons. The correction F to the electron backscattering was estimated numerically [196].

The beta-emission asymmetry was studied on the ITEP apparatus with the ^{8}Li, ^{20}F, ^{66}Cu, ^{108}Ag, and ^{110}Ag nuclei. The spins of excited states formed by thermal neutron capture were determined, and the ground state magnetic moments of ^{8}Li and ^{20}F nuclei were measured [186].

Apparatuses similar to that described above were also constructed at the TRIGA [197] and FR2 reactors in Karlsruhe [188, 198], and at the FRM reactor in Munich [199]. However, polarized neutron beams in the studies [188, 198] were formed by the diffraction reflection from ferromagnetic crystals, so that the polarized beam intensities were low. A five-meter polarizing neutron guide was used in [199] (see entry 5 in Table 1.1).

Rauch [197] studied the asymmetry of the beta emission from ^{116}In; however, doubts were later expressed regarding the correctness of the reported results [188]. The magnetic moments of the ^{116}In and ^{110}Ag nuclei were measured in [188, 198], the quadrupole moments of the ^{8}Li, ^{20}F, and ^{116}In nuclei were measured in [188, 190, 200], and the magnetic moments of the ^{28}Al, ^{38}Cl, and ^{116}In nuclei were reported in [199]. Later, the neutron polarizer at the FR2 reactor was replaced by a magnetized CoFe mirror, and then by a multislit collimator consisting of permendur mirrors, whose characteristics are given in Table 1.1 (entry 16). A system based on an NMR spectrometer was designed at the high-flux ILL reactor in Grenoble [201] to measure the magnetic moments of the ^{108}Ag and ^{110}Ag nuclei.

The processes of generation and anneal of point defects appearing in various single crystals as a result of the recoil of nuclei, excited in (n,γ)-nuclear reactions by the emission of gamma rays, were studied in a number of recent papers [202]; see also [187, 203, 204]. An analysis of these effects provides information on hyperfine interactions between beta-radioactive nuclei and defects, defect anneal rates, etc.

The ITEP apparatus detected the multispin magnetic resonance involving polarized beta-radioactive nuclei [205]. The beta-radioactive ^{8}Li nuclei in the LiF specimen underwent resonance depolarization when the specimen was subjected to an rf field not only at the Larmor frequency of ^{8}Li nuclei but also at frequencies which were linear combinations of the Larmor frequencies of the ^{8}Li, ^{7}Li, and ^{19}F nuclei with integral coefficients.

9.3 Summary of Experimental Data

The experimental results are summarized in Table 9.2. The w_+ column gives the lower bound on the probability for the thermal neutron capture by nuclei of going through the $J_c = J_i + 1/2$ channel. The nuclear magnetic moments are given without correction for atomic diamagnetism. These corrections are below 0.6 %. The quadrupole nuclear moments are given without polarization corrections.

10. Anisotropy of Alpha Particles and Other Light Nuclei Emitted After Polarized Neutron Capture

In 1977 *Lobov* and *Danilyan* [206] suggested a novel approach to studying the violation of spatial parity in nuclear interactions; namely, an analysis of the anisotropy of alpha particles and other light nuclei emitted after a nucleus captures a polarized neutron. Two reactions were discussed:

$$^6\text{Li}(\vec{n}, t)^4\text{He} \; ; \quad \text{and} \tag{10.1}$$

$$^{10}\text{B}(\vec{n}, \alpha)^7\text{Li} \; . \tag{10.2}$$

In addition, the reaction

$$^3\text{He}(\vec{n}, p)^3\text{H} \tag{10.3}$$

was studied in [207]. All these reactions are of considerable interest because they involve light nuclei which permit a more detailed theoretical analysis than heavy and medium nuclei do.

The experiments were first carried out by *Lobashev* and his coworkers at LINP in 1979 [208]. The difficulty was to ensure the statistical error of measurements at the level of 10^{-7}–10^{-5}. Particle counting failed at high counting rates, so that the method used in the experiments [207, 208] was the integral measurement of high particle fluxes; it was developed earlier at LINP for measuring low circular polarization of gamma quanta emitted by nonpolarized nuclei [209]. The effects due to light and heavy reaction products, for example, alpha particles and ^{7}Li nuclei in the reaction (10.2), were separated on the basis of the difference in free path lengths in the gas between the target and the detecting volume.

However, considerable P-even left-right asymmetry was observed in the reaction (10.1) already in the first LINP study [208]. The same asymmetry was later found in the reaction (10.2) [207] for alpha particles connected with the formation of the recoil ^{7}Li nucleus in the ground state (the α_0 transition). Such left-right asymmetry imposes a limit on the measurable P-odd anisotropy of the alpha emission because it produces a spurious effect. For this reason, a special experiment geometry was suggested at ITEP, which substantially suppressed the effect of left-right asymmetry [210].

10.1 Fundamentals of the Experimental Method

The nucleon-nucleon weak interaction in nuclei [which was established in a number of experiments including the study of the anisotropy of gamma rays emitted by nuclei after polarized neutron capture (see Chap. 8)] mixes states with identical spin but opposite parities. Consequently, if a nucleus is polarized and emits alpha particles (or other light nuclei), the selection rules for momentum and parity make the emission of alpha particles (or other light nuclei) with orbital moments L or $L' = L + 1$ possible.

This behavior results in an angular correlation of alpha particles:

$$w(\theta) \propto 1 + a(\boldsymbol{s}\boldsymbol{p}_l) = \text{const}(1 + P_n a \cos \theta) \ , \tag{10.4}$$

where a is the coefficient of the P-odd symmetry, $\boldsymbol{s}$ is the neutron spin, $\boldsymbol{p}_l$ is the momentum of the light particle, P_n is the neutron beam polarization, and θ is the angle between the neutron spin and the alpha-particle momentum. This correlation is similar to the correlation in gamma emission from nuclei after polarized neutron capture, taking into account the nucleon-nucleon weak interaction (see Sect. 8.1.2). The P-odd asymmetry coefficient a is found from (10.4), using a formula similar to (8.2).

According to [206], the expression for the asymmetry coefficient a includes the quantity $\Delta_{LL'} = |A_{L'}|/|A_L|$, which is the ratio of the modules of probability amplitudes for the emission of alpha particles with orbital momenta L and L'. The probability of emission of an alpha particle with admixed orbital momentum L' is nonzero because of the nucleon-nucleon weak interaction in the nucleus. For this reason, $\Delta_{LL'}$ is proportional to the weak coupling constant and is of the order of 10^{-6}. However, if $L' = L - 1$, the P-odd effect is kinematically enhanced because of a lower centrifugal barrier $\hbar^2 L'(L'+1)/2mR^2$ for a nonregular alpha transition than in the case of a centrifugal barrier $\hbar^2 L(L+1)/2mR^2$ for a regular alpha transition (m is the alpha particle mass, and R is the nuclear radius). Contrariwise, the effect is reduced if $L' = L + 1$.

P-odd effects also depend on the phase of interference of the competing spin channels and on the relative phase of interference of the two states with opposite parities in each of the spin channels. However, these dependences are rather weak [206].

At the same time, a left-right asymmetry, conserving the spatial parity, was found in the reactions (10.1) and (10.2) for the emission of light particles. The asymmetry is caused by the interference of s- and p-waves in the amplitudes of the (n, α)- and (n, t)-reactions. In this case, the mixing of the levels of the compound nucleus results not from the weak interaction but from the overlapping of the s- and p-wave neutron resonances [211]. When a neutron is captured from the s- and p-wave states, the resultant compound-nucleus states have opposite parities. The observed P-even correlation has the form

$$w(\theta) \propto 1 + a_{LR}\boldsymbol{s}(\boldsymbol{p}_n \times \boldsymbol{p}_l) \ , \tag{10.5}$$

where a_{LR} is the left-right asymmetry coefficient; s, p_n, p_l are unit vectors in the direction of the neutron spin, its momentum, and the momentum of the light particle, respectively. The term $s(p_n \times p_l)$ is P-even and T-noninvariant (see Table 6.1). However, the existence of this term does not mean that the T-invariance is broken in the reactions mentioned. The coefficient a_{LR} contains a factor $\sin(\delta_s - \delta_p)$, where $\delta_{s(p)}$ is the phase of neutron capture by a nucleus in the $s(p)$ state. This factor also reverses its sign under time reversal [212], so that on the whole, the correlation (10.5) is T-invariant. It reaches a maximum when all three vectors s, p_n, p_l are mutually orthogonal.

10.2 Survey of Experiments

10.2.1 Selection of Nuclei

Reactions (10.1-3) were chosen for study in view of their relatively large cross sections. Let us analyze the reaction (10.2) in more detail. The scheme of nuclear levels participating in the reaction is shown in Fig. 10.1. The cross section of the reaction (n, α) on thermal neutrons and the ^{10}B nucleus is large (about 3840×10^{-24} cm^2). The spin and parity of the ground state of ^{10}B are $J_i^\pi = 3^+$. The final state of the nucleus ^{7}Li*, populated at a probability of 93.3%, has a spin and parity $J_f^\pi = 1/2^-$ and decays into the ground state of ^{7}Li and a gamma quantum of energy $E = 0.478$ MeV. The energy of alpha particles is $E = 1.478$ MeV (the α_1 transition). In 6.7% of the events, the compound nucleus ^{11}B decays directly into the ground state of the ^{7}Li nucleus, emitting a 1.786 MeV alpha particle (the α_0 transition).

It was shown [213] that the α_1 transition predominantly (i.e., by more than 96%) goes through the spin state $J_c^\pi = 7/2^+$ of the ^{11}B nucleus. The magnitude of the P-odd effect is a function of the final states. The orbital momentum carried by the alpha particle in the transition from the state with spin J_c to the state with spin J_f follows from the inequality $|J_c - J_f| \leq L \leq J_c + J_f$, and the parity of the final state, π_f, is related to the initial state parity, π_i, and the parity of the alpha particle, $\pi_\alpha = +1$, by the formula $\pi_f = \pi_i \pi_\alpha (-1)^L$. Consequently,

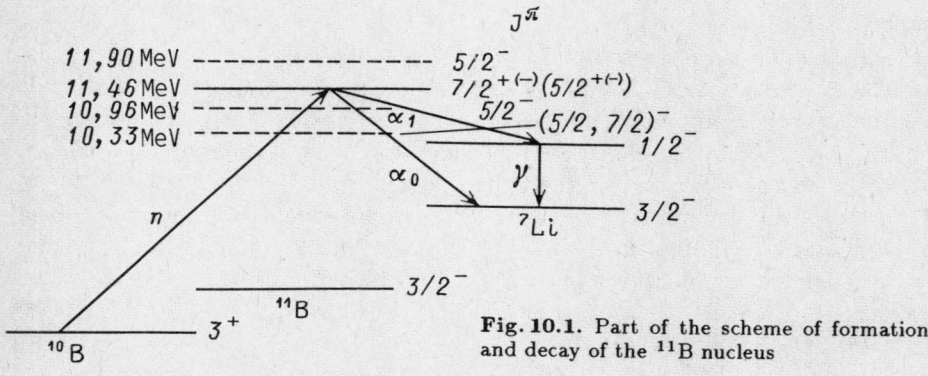

Fig. 10.1. Part of the scheme of formation and decay of the ^{11}B nucleus

if $J_c^\pi = 7/2^+$ and $J_f^\pi = 1/2^-$, the parity-allowed orbital momentum carried by the alpha particle is $L = 3$, and the admixed (nonregular) orbital momentum is $L' = 4$. Therefore, in this case (the α_1 transition), the effect is reduced because of a higher centrifugal barrier for the nonregular transition (see Sect. 10.1). If, however, $J_c^\pi = 5/2^+$ and $J_f^\pi = 3/2^-$, the allowed orbital momenta of alpha particles are $L = 1$ and $L = 3$, and the admixed orbital momenta are $L' = 2$ and $L' = 4$. In this case (the α_0 transition), the effect may be enhanced. Furthermore, one has to take into account that in this case, the scheme of the levels of ^{11}B contains 11.90 and 10.96 MeV levels with spin and parity $5/2^-$, close to the excitation energy 11.46 MeV of the compound nucleus; hence, the P-odd effect results from the overlapping of the levels with identical spins and opposite parities ($5/2^+$ and $5/2^-$).

10.2.2 Requirements for the Detector Unit of the Apparatus

The expected asymmetry of the emitted particles in the reactions studied does not exceed 10^{-7}–10^{-5}, so that it is necessary that the detecting units of the system satisfy several important and frequently contradictory requirements:

1. The target must be sufficiently thick to ensure maximum absorption of the neutron beam;
2. charged reaction products must escape from the target without losing energy; that is, the target must be sufficiently thin;
3. the intensity fluctuations of the neutron beam must be compensated for; thus, it is advisable to have two detecting units to simultaneously record the reaction products escaping along and against the direction of the neutron spins [at $\theta = 0$ and $\theta = 180°$, respectively, in (10.4)];
4. maximum possible particle fluxes must be recorded, so that the integral method of measurements is advisable;
5. the effects due to light and heavy reaction products must be separated, preferably by employing the difference between free path lengths of light and heavy reaction products in gases.

10.2.3 Experimental Systems

The system developed at LINP [207, 208] (Fig. 10.2) satisfies the requirements listed above. A thin target was placed at a small angle to the incident neutron beam with the detector units on both sides of the target. Measurements were carried out with the LINP thermal transversely polarized beam, at the 6×10^7 neutron/s flux.

The direction of the neutron beam was perpendicular to the plane in Fig. 10.2. Multifilament ionization chambers, operating in the proportional mode, were used as detector units. The targets were boron and lithium layers enriched in the ^{10}B and ^{6}Li isotopes, coated with titanium layers on both sides. In working with ^{3}He, the target was that part of the chamber filled with a ^{3}He(98 %)+CO_2(2 %) mixture through which the beam passes. In studying

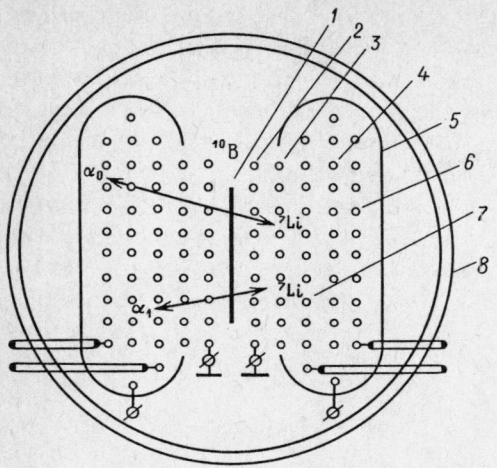

Fig. 10.2. The LINP detector for studying the anisotropy of alpha emission in the reaction $^{10}\text{B}(n,\alpha)^{7}\text{Li}$ [208]: *(1)* ^{10}B target, *(2)* grounded grid, *(3–5)* high voltage grid and solid electrodes, *(6.7)* signal electrodes of the outer and inner gaps connected with preamplifier inputs, *(8)* chamber housing

the P-odd asymmetry, the neutron spins were oriented along the horizontal axis [yielding the maximum of the correlation (10.4)]; in studying the left-right asymmetry, they pointed along the vertical axis [yielding the maximum of the correlation (10.5)].

The ions produced in the working gaps of the chamber between the high-voltage electrodes were collected by the output signal electrodes. The gas pressure in the chamber was so adjusted that the heavy reaction products did not reach the working gap. When the reaction (10.2) was studied, the gap was additionially partitioned into outer and inner parts (see Fig. 10.2), so that the outer gaps detected only the alpha particles of the α_0 transition, while the inner gaps recorded the alpha particles of both transitions. The current output of signal electrodes was converted into voltage in preamplifiers, amplified, and then fed into the differential amplifier which subtracted the current output by the opposite gaps. As a result, asymmetry effects in the current difference added up, while the neutron flux fluctuations tended to cancel out.

The neutron polarization vector was reversed with respect to the leading magnetic field direction by an adiabatic rf spin flipper (see Sect. 1.2.3) at a period of 4 s.

The asymmetry coefficient yielded by a single measurement was found as follows:

$$A = \frac{(I_L - I_R)_+ - (I_L - I_R)_-}{(I_L + I_R)_+ + (I_L + I_R)_-}, \qquad (10.6)$$

where $I_{L(R)}$ is the current in the left (right) gap averaged over the duration of a single measurement, and the plus (minus) subscript corresponds to the direction of neutron spins along (against) the field; this correspondence is imposed by switching off (on) the rf field of the spin flipper.

The effect of the magnetic field of the spin flipper on the sensitivity of the detecting unit was eliminated by reversing the permanent magnetic field on the detector every 12 hours. The direction of the neutron polarization was reversed together with the field ("+ field" and "− field"). The sign of the sought effect was thereby reversed, while the sign of the possible spurious effect remained unchanged.

The asymmetry coefficient in the daily cycle was found by calculating

$$a = \frac{A(+\text{ field}) - A(-\text{ field})}{2P_n \overline{\cos\theta}}, \qquad (10.7)$$

where $P_n = 0.97$ is the neutron beam polarization, and $\overline{\cos\theta}$ is the mean value of $\cos\theta$ given by (10.4); $\overline{\cos\theta}$ is determined by the geometry of the target and of the working gaps.

Large left-right asymmetry effects in the reaction (10.1) and in the α_0-transition for alpha particles in the reaction (10.2) were found in the LINP system. The results are shown in Table 10.1 (see below).

Table 10.1. The left-right asymmetry coefficients a_{LR} obtained in the LINP experiments

Reaction	a_{LR}, $[10^{-6}]$	Ref.
$^3\text{He}(\vec{n},p)^3\text{H}$	-0.34 ± 0.57	[207]
$^6\text{Li}(\vec{n},t)^4\text{He}$	95 ± 4	[208]
$^{10}\text{B}(\vec{n},\alpha_0)^7\text{Li}$	77 ± 6	[207]
$^{10}\text{B}(\vec{n},\alpha_1)^7\text{Li}$	-2.8 ± 1.4	[207]

As a result of the left-right asymmetry, it was not possible to improve the accuracy of the experiments on the detection of the P-odd asymmetry: The left-right asymmetry effects contributed a systematic error to the results of the P-odd asymmetry measurements. This complication resulted in an additional requirement for the detecting units of the system: The neutron beam polarization vector must point along or against the neutron momentum. In this case, the term $s(p_n \times p_l)$ in (10.5) vanishes. It is then necessary to analyze the asymmetry of emission of the reaction products in, as well as counter to, the direction of polarization of the longitudinally polarized neutron beam.

This constraint was met in the joint ITEP-ECN experiment carried out with a polarized neutron beam of the Petten high flux reactor [210, 214].

The ITEP-ECN detector had a target divided into thin layers, each of which was placed in a detection module (Fig. 10.3). The target in a module consisted of a 170 μg/cm^2 boron layer, enriched in ^{10}B (85 %), which was deposited on a titanium substrate (50 μg/cm^2) and covered by a similar titanium layer. Two sensitive gaps were arranged at the sides of the target. Each gap was a multiwire ionization chamber.

The detector contained 12 such modules placed one after the other. It was irradiated with a longitudinally polarized beam of thermal neutrons (entry 6 of Table 1.1) at a flux of 3×10^8 neutron/s, polarized to $P_n = 0.9$. The size of

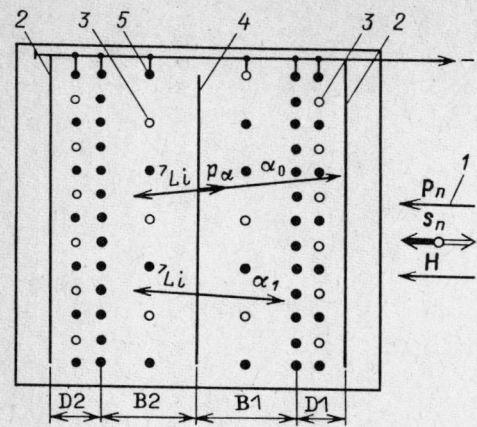

Fig. 10.3. The ITEP-ECN detector module for studying the anisotropy of alpha emission in the reaction $^{10}B(n,\alpha)^{7}Li$ [210]: *(1)* longitudinally polarized neutron beam, *(2)* boundary electrodes, *(3)* signal wire electrodes, *(4)* ^{10}B target, *(5)* high-voltage wire electrodes, *(D1,D2)* farther gaps, *(B1,B2)* closer gaps

the detector window was 62×110 mm. The electrodes were wound with gold-coated tungsten wire, 20 μm in diameter, at steps of 2 and 4 mm. In addition to wire electrodes, each module contained one target electrode and two boundary electrodes of aluminum foil, separating the modules. Information was read off the wire electrodes in the integral mode; i.e., ionization currents [caused by the escape of products of reaction (10.2) from the targets] were recorded. The selection of the α_1 or α_0 transition was effected by adjusting the argon pressure in the detector. Consequently, the ionization currents in the gaps far from the targets ($D1$ and $D2$) were mostly caused by the alpha particles of the analyzed transition, and those in the gaps closest to the targets ($B1$ and $B2$) were caused both by alpha particles and by ^{7}Li ions. Furthermore, owing to the natural collimation of the particle flux, the angle of the cone through which alpha particles enter gaps $D1$ and $D2$ is substantially smaller than the corresponding angle for gaps $B1$ and $B2$. Therefore, the anticipated P-odd effect in gaps $D1$ and $D2$ was greater than that in gaps $B1$ and $B2$ by a factor of three, making it possible to use $B1$ and $B2$ for control measurements.

The neutron beam polarization vector was reversed with periods of 0.85 and 3.28 s, chosen in view of best suppression of the reactor noise contribution after an analysis of the correlations in the frequency spectra of the reactor and detector noise signals. At the end of each period, the currents, read off all four detector gaps, were converted into a digital code and fed to the data storage system; every 1023 periods, the data were written into a magnetic tape.

The direction of the longitudinal magnetic field was reversed three times a day. A control experiment with a depolarized neutron beam was also run. The asymmetry coefficient was calculated with formulas similar to (10.6) and (10.7).

The LINP experiment was repeated [215] using a detector whose geometry was similar to that described in [214]. The results of these two experiments are given in Table 10.2.

Table 10.2. Upper bounds of the asymmetry coefficients

Reaction	a, $[10^{-6}]$	Source
$^3\text{He}(\vec{n},p)^3\text{H}$	<1.2	LINP [207]
$^6\text{Li}(\vec{n},t)^4\text{He}$	<1.4	LINP [215]
$^{10}\text{B}(\vec{n},\alpha_0)^7\text{Li}$	<8.0	LINP [215]
	<3.7	ITEP and ECN [214]
$^{10}\text{B}(\vec{n},\alpha_1)^7\text{Li}$	<1.5	LINP [215]
	<0.6	ITEP and ECN [214]

Note. The confidence level for the coefficient a is 90%

10.3 Summary of Results

The left-right asymmetry coefficients are given in Table 10.1.

The upper bounds of the P-odd asymmetry coefficients are listed in Table 10.2.

As follows from Table 10.2, the P-odd asymmetry of the emission of alpha particles and other light nuclei has not been detected in the selected reactions. Its absence in the reaction $^{10}\text{B}(n,\alpha_0)^7\text{Li}$ makes one assume either that the transition α_0 is also connected with the spin channel $J_c^\pi = 7/2^+$ of the compound nucleus ^{11}B, or that an unfavorable combination of orbital momenta of alpha particles, L and L', is realized for the spin channel $J_c^\pi = 5/2^+$; this combination does not result in the enhancement of the P-odd effect because the centrifugal barrier for the nonregular α transition is not lower than that for the regular α transition. In any case, the upper bound of the matrix element of the weak interaction which mixes the opposite-parity states in the compound nucleus ^{11}B is not greater than 2 eV.

Recently, a P-odd effect was reported for the first time [216] in the reaction in which light nuclei are emitted after polarized neutron capture; namely, the P-odd asymmetry of proton emission in the reaction $^{35}\text{Cl}(\vec{n},p)^{35}\text{S}$. The experiment was carried out with the LINP transversely-polarized thermal neutron beam, in the geometry identical to that of [207, 208]. The P-odd asymmetry coefficient was found to be rather large: $a = (-1.51\pm0.34)\times10^{-4}$, while the left-right asymmetry coefficient was $a_{\text{LR}} = -(2.40\pm0.43)\times10^{-4}$.

11. Anisotropy of the Angular Distribution of Fragments After Fission of Heavy Nuclei by Polarized Neutrons

In 1961, *Vladimirsky* and *Andreev* [217] came up with a hypothesis of a possible anisotropy of emission of light (and heavy) fragments due to the nonconservation of parity in the spontaneous fission of heavy polarized nuclei. This hypothesis was based on the argument that the fission threshold may be a function of the spin and parity of the fissioning nuclei [218]. If the barrier for fission from the state with nonregular parity is lower than that from the state with regular parity, the barrier enhancement of asymmetry is possible in the emission of light (and heavy) fragments along and against the direction of spin of the fissioning nucleus. No such experiment has been carried out so far, due to the extreme complexity involved.

The possibility of detecting the effect of mixing of states with opposite parities in the two-hump model of subbarrier fission of some nuclei (e.g., ^{240}Pu) by polarized resonance neutrons was discussed in [219]. These experiments were not realized either.

Only in 1976 did *Danylyan* et al. [220] conduct an experiment at ITEP on the P-odd asymmetry of fragment emission from ^{235}U nuclei fissioned by polarized thermal neutrons. This asymmetry is given by the formula

$$w(\theta) = \text{const}(1 + a\boldsymbol{s}\boldsymbol{p}_l) = \text{const}(1 + P_n a \cos\theta) \;, \tag{11.1}$$

where a is the asymmetry coefficient, $\boldsymbol{s}$ is the unit vector in the direction of the nuclear spin, $\boldsymbol{p}_l$ is the unit vector in the direction of the momentum of the light fragment, θ is the angle between $\boldsymbol{s}$ and $\boldsymbol{p}_l$, and P_n is the neutron beam polarization.

Danilyan et al. [220] used the following arguments [221]. If, in the course of fission, the nucleus does not "forget" (contrary to usual assumptions) its compound state, then the dynamic enhancement mechanism responsible for the appreciable asymmetry of gamma emission in radiative neutron capture (see Sect. 6.6) is also realized in the fission channel. The level density of the fissioning compound nuclei is greater by roughly an order of magnitude than that of the compound ^{114}Cd and ^{118}Sn nuclei. We can expect, therefore, that the fissioning compound nuclei also contain a substantial admixture of states with the opposite parity. Nuclear fission differs from the emission of gamma quanta by a compound nucleus in that the number of allowed final states is much greater in the former case than in the latter, ranging from 10^7 to 10^{10}.

As shown in Sect. 8.2.2 for the ^{114}Cd nucleus, the magnitude and sign of the asymmetry coefficient a depend on the quantum characteristics of the final state; hence, one has to expect a statistical averaging of asymmetry coefficients unless a definite final state is singled out. In fact, this procedure is not feasible, so the resultant asymmetry would be close to zero.

The above analysis may be incorrect, owing to some specifics of fission. When some heavy nuclei are fissioned by thermal neutrons, the fragments are known to have unequal masses. The most natural interpretation of this observation is to assume that fragment formation is preceded by asymmetrically deformed nuclear states. If the weak interaction mixes such states with opposite parities, the fragment emission asymmetry is independent of the final state. The fission process goes through a small number of intermediate states (fission channels) characterized by certain values of a quantum number K (nuclear spin projection on the deformation axis) [222]. In this case, averaging over a small number of asymmetrically deformed intermediate states having opposite parities may result in a nonzero asymmetry coefficient in the emission of light (and heavy) fragments.

11.1 Survey of Experiments

11.1.1 Specific Features of Fission Reactions

Methodologically, the experiments searching for the P-odd asymmetry in the emission of fission fragments are less complicated than those aimed at studying the asymmetry in (n,γ)-, (n,β)- or (n,α)-reactions. Fission is specific in that the asymmetry in the emission of the "particles", i.e., lighter fragments, and the emission of "recoil nuclei"; i.e., heavy fragments, can be measured simultaneously in the same experiment; moreover, these asymmetry coefficients must have equal magnitudes and opposite signs. This constraint gives an additional criterion of reliability to the experimental data. Almost no difficulties are encountered in background discrimination and in the overlapping of the energies of the lighter and heavier fragments.

11.1.2 Discovery of P-Odd Asymmetry in the Emission of Fission Fragments

We shall now discuss the ITEP experiment which revealed the asymmetry in the emission of fission fragments (Fig. 11.1) [220]. The parameters of the polarized neutron beam used in this experiment have been given in Chap. 1 (entry 1 in Table 1.1). The neutron spins were oriented in the horizontal plane. The target consisted of five aluminum disks, 30 mm in diameter and 0.1 mm thick, coated on both sides by a 100 μg/cm^2 layer of uranium oxide enriched in ^{235}U (75 %). The disks were mounted along the beam axis. The fragments ejected from the target were detected by silicon surface-barrier detectors placed on both sides

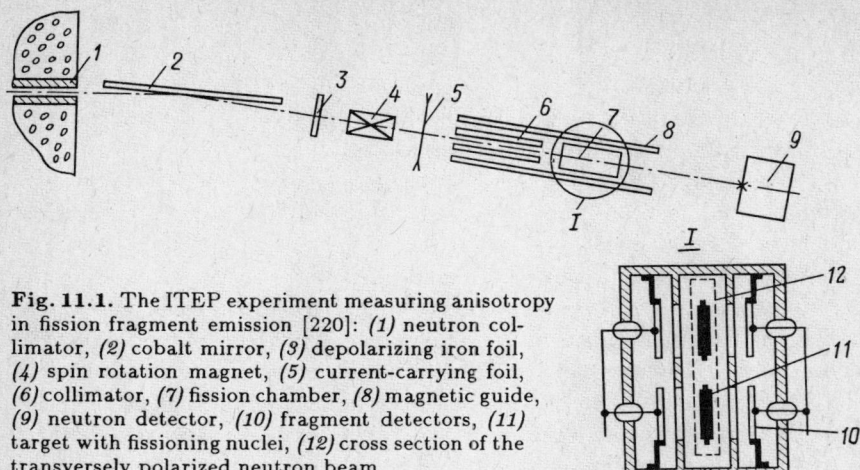

Fig. 11.1. The ITEP experiment measuring anisotropy in fission fragment emission [220]: *(1)* neutron collimator, *(2)* cobalt mirror, *(3)* depolarizing iron foil, *(4)* spin rotation magnet, *(5)* current-carrying foil, *(6)* collimator, *(7)* fission chamber, *(8)* magnetic guide, *(9)* neutron detector, *(10)* fragment detectors, *(11)* target with fissioning nuclei, *(12)* cross section of the transversely polarized neutron beam

of each disk. Electronic circuits selected the pulses corresponding to the groups of light and heavy fragments, and fed these pulses to different scalers. The beam polarization was flipped once a second. The asymmetry coefficient for light (heavy) fragments was calculated using the formula

$$a_{l(h)} = \left(\frac{N_+ - N_-}{N_+ + N_-}\right)_{l(h)}, \qquad (11.2)$$

where $N_{+(-)}$ is the counting rate of light (heavy) fragments for two opposite orientations of neutron spins. The resulting asymmetry coefficient a was found as the arithmetic mean of the coefficients a_l and a_h.

Special measures were taken to avoid an instrument-caused asymmetry; namely an asymmetry due to nonsimultaneous measurements of fragment counting rates for two opposite beam polarization directions. Correspondingly, the apparatus was symmetrized, as it was for the reactions (n, γ), (n, β), and (n, α); i.e., two independent target-detector systems were arranged symmetrically with respect to the neutron beam. In this geometry, the asymmetry due to slow variations of the neutron beam intensity and the parameters of electronic circuits was eliminated by flipping the beam polarization. Measurements with a depolarized beam were carried out as control experiments.

Four months of 24-hours-a-day measurements yielded the following asymmetry coefficient of light fragment emission in the fission of ^{236}U:

$$a(^{236}\text{U}) = (1.37 \pm 0.35) \times 10^{-4}.$$

The plus sign of the asymmetry coefficient signifies that lighter fragments are preferably emitted in the direction of neutron spin orientation.

The second experiment carried out by the same group [223] analyzed the asymmetry in the emission of fission fragments from ^{239}Pu nuclei. The arrangement of detectors with respect to the target was altered so as to simultaneously measure the P-odd correlation sp_l and the P-even correlation $s(p_n \times p_l)$, where p_n is the neutron momentum. This correlation is possible in the interference of s- and p-waves in the course of neutron capture. The correlation was later experimentally discovered (see Sect. 11.3). In an ideal geometry, no such correlation arises if the vector s is parallel to the vector p_l, but the deviation of the actual geometry from the ideal one could result in its appearance. It was shown that the contribution of the P-even correlation to the P-odd one did not exceed 4%. The asymmetry coefficient for ^{240}Pu had the sign opposite to that of ^{236}U:

$$a(^{240}\text{Pu}) = -(4.8 \pm 0.8) \times 10^{-4} \ .$$

Finally, the third work of the same group [224] studied the asymmetry in the fragment emission from ^{233}U nuclei. Their analysis yielded the value

$$a(^{234}\text{U}) = (2.8 \pm 0.3) \times 10^{-4} \ .$$

Immediately after the results of the ITEP experiments were published, similar work was begun at other institutes. The asymmetry in the emission of lighter fragments from fissioning ^{235}U and ^{233}U nuclei, irradiated by polarized thermal neutrons, was measured at LINP [208, 225]. Fragments were recorded by the integral technique developed earlier at this institute for measuring the circular polarization of gamma quanta emitted by nonpolarized nuclei [209]. The authors made use of the fact that the free path length of light fragments in materials is slightly greater than that of heavy fragments. It was therefore possible to adjust the pressure of the gas in the gap between the target and the detector so that only the light fragments could reach the sensitive region of the detector (i.e., the multiwire proportional chamber, some distance away from the target).

Later, the ITEP group repeated the measurements of the asymmetry coefficient for three fissioning nuclei, employing fission ionization chambers instead of semiconductor detectors for recording the fission fragments [226]. The results of all measurements are summarized in Table 11.1 below.

The P-odd asymmetry coefficient for the emission of fission fragments by ^{241}Pu nuclei irradiated by thermal polarized neutrons was measured in [227]. Up to experimental error, no asymmetry was found.

11.1.3 An Analysis of the Asymmetry Coefficient as a Function of Fragment Mass and Neutron Energy

It would be interesting to find out whether the asymmetry in the emission of fission fragments is influenced by the fragment mass. An attempt to establish this effect was first made in [224]. The asymmetry coefficient was measured separately for the groups of light and heavy fragments in four intervals of the amplitude spectrum of ^{233}U fission fragments; i.e., each fragment group was

Table 11.1. Asymmetry coefficients a for fissioin fragment emission

Compound nucleus	J_i^π	a, $[10^{-4}]$	Source
^{234}U	$5/2^+$	2.8 ± 0.3	ITEP [224]
		3.8 ± 0.6^a	ITEP [234]
		3.6 ± 0.3	LINP [225]
		4.2 ± 0.3	ITEP [226]
		4.4 ± 0.1	LINP [227]
^{236}U	$7/2^-$	1.37 ± 0.35	ITEP [220]
		0.84 ± 0.06	LINP [208]
		$0.5\ \pm0.3^a$	ITEP [234]
		0.75 ± 0.12	LINP [225]
		$1.1\ \pm0.2$	ITEP [226]
^{240}Pu	$1/2^+$	$-4.8\ \pm0.8$	ITEP [223]
		$-7.8\ \pm0.8^a$	ITEP [234]
		-6.22 ± 0.35	LINP [228]
		$-6.7\ \pm0.9$	ITEP [226]
^{242}Pu	$5/2^+$	-0.27 ± 0.29	LINP [227]

^{a} The result is obtained by recalculation of the asymmetry coefficient for fission neutron emission

divided into two parts, each referring to different energies. No dependence of the asymmetry coefficient on fragment mass was observed.

A more detailed study was carried out at LINP [228, 229] for ^{233}U and ^{239}Pu nuclei. Light and heavy fragments escaping from a thin target in opposite directions along the axis of neutron spin orientation were recorded by two semiconductor detectors in the coincidence circuit. The distribution of the ratio $E_1/(E_1 + E_2)$ was plotted, where $E_{1,2}$ stands for the pulse amplitudes of the first and second detectors, proportional to the fragment's kinetic energy. These ratios characterized the mass of a fragment detected by the second detector (Fig. 11.2). The solid curve shows the measured mass spectrum of the fragments. The figure demonstrates that the asymmetry coefficient is independent of the fragment mass.

The asymmetry in the emission of fission fragments by ^{239}Pu irradiated by polarized neutrons in the 0.01–0.7 eV energy range was studied in the experiment [230]. Close to the resonance, at $E = 0.297$ eV, the asymmetry coefficient was $-(13.9\pm3.4)\times10^{-4}$, i.e., approximately twice as large as the coefficient in the fission of ^{239}Pu irradiated by thermal neutrons.

The asymmetry coefficient of emission of fission fragments from ^{233}U was studied in [231] as a function of energy in the vicinity of the expected p-resonance in the ^{233}U cross section at 0.17 eV. It was shown that the P-odd asymmetry coefficient reverses its sign in the resonance region. Consequently, the transmission of ^{233}U would depend appreciably on the helicity of neutrons, owing to the interference of s- and p-wave capture states (see Sect. 12.1) in the range of thermal energies [126]. An analysis of the P-odd asymmetry of the total

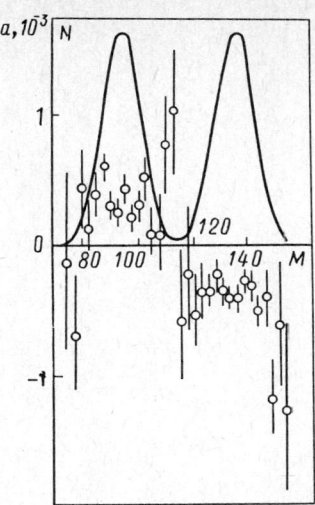

Fig. 11.2. Asymmetry coefficient as a function of fragment mass in the fission of ^{233}U nuclei induced by polarized neutrons [228]. The solid curve $N(M)$ is the spectrum of fragment masses

interaction cross section between thermal longitudinally polarized neutrons and ^{233}U nuclei [232, 233] demonstrated the absence of the sought asymmetry effect in the total cross section at a level which makes the hypothetical p-wave nature of the resonance at 0.17 eV doubtful.

11.1.4 Asymmetry in Fission Neutron Emission

Other experiments, which independently confirmed the anisotropy in the emission of light fragments, were carried out at ITEP [234], measuring the asymmetry in the emission of fission neutrons. Fission neutrons are known to be emitted mostly along the momentum of light fragments [235]; hence, the anisotropy in the emission of fission neutrons could also be expected. This method permits work with sufficiently thick targets of fissioning nuclei ^{233}U, ^{235}U, and ^{239}Pu, which absorb most of the incident-polarized neutrons. As a result, the counting statistics is built up faster than in studying the asymmetry in fragment emission, although the sensitivity of this technique to the asymmetry in fragment emission is lower, as is the measured effect.

The experiment was carried out with the same polarized neutron beam used in studying the anisotropy of gamma emission after polarized neutron capture (see entry 3 in Table 1.1). Fission neutrons were detected by two plastic scintillators in the directions along the neutron beam polarization vector and against it. The direction of polarization was flipped at a frequency of 8 Hz. Lead filters were used to reduce the background caused by gamma-ray emission from the target. A control experiment was run with a depolarized beam. A series of additional control experiments demonstrated that the beam displacement caused by flipping the beam polarization, the fluctuations of beam intensity, and electronic instability were within the limits ensuring an asymmetry measurement error of less than 10^{-5}.

Calibration experiments, using a thick target of fissioning nuclei replaced by a thin one and semiconductor detectors to detect fission fragments, were conducted for a quantitative comparison of asymmetry coefficients in the emission of neutrons and fragments of fission. Coincidence counts were recorded for plastic scintillator pulses and pulses due to light and heavy fragments for different angles between the fragment emission axis and the direction toward the neutron detector.

The following values of the asymmetry coefficients in the emission of fission neutrons were reported [234]:

$$a_n(^{234}U) = (4.5 \pm 0.7) \times 10^{-5},$$

$$a_n(^{236}U) = (0.7 \pm 0.4) \times 10^{-5},$$

$$a_n(^{240}Pu) = -(6.7 \pm 0.7) \times 10^{-5}.$$

In Table 11.1, these values are recalculated into the asymmetry coefficients of fission fragment emission, using the results of calibration experiments.

11.2 Summary of Experimental Data

The available results are summarized in Table 11.1. The tabulated asymmetry coefficients are corrected for the finite solid angle in which fragments are detected; they are normalized to 100 % neutron beam polarization, but not to 100 % nuclear polarization. Difficulties arise in the interpretation of the reported results because the compound states of the fissioning nuclei are a mixture of two spin channels $J_c = J_i \pm 1/2$, with different magnitudes and signs of nuclear polarization. Consequently, the sign of the asymmetry coefficient is given conditionally; namely, relative to the orientation of neutron spins.

Although the results obtained with different systems differ by a factor of 1.5–2 beyond the experimental error (probably pointing to a number of unidentified systematic effects), the general picture appears to be fairly self-consistent: the asymmetry is observed quite reliably, the sign of a for the ^{240}Pu nucleus is the opposite of the sign for the ^{234}U and ^{236}U nuclei, and there presumably is a correlation between the asymmetry coefficient and the spin of the target nucleus.

11.3 Investigation of P-Even Angular Correlations

Experiments at LINP revealed [225. 228] that the violation of spatial parity in the fission of ^{233}U and ^{235}U is accompanied by a left-right asymmetry (conserving the spatial parity) in the emission of fission fragments; this asymmetry was caused by the interference of s- and p-wave fission resonances. This obser-

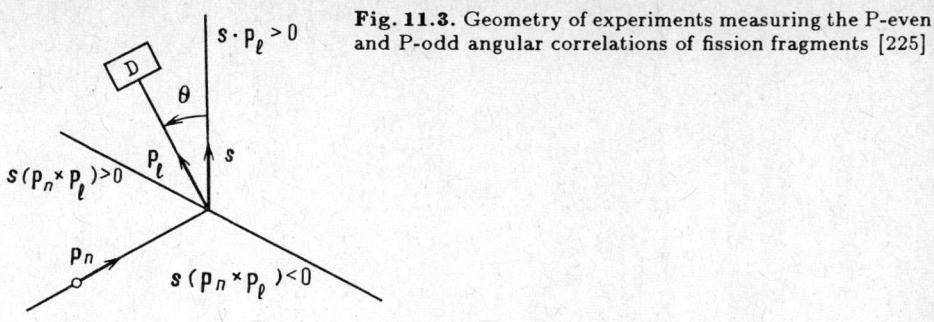

Fig. 11.3. Geometry of experiments measuring the P-even and P-odd angular correlations of fission fragments [225]

vation is similar to the interference of s- and p-waves in the reactions (n, α) and (n, t) (see Sect. 10.1).

Two multiwire proportional chambers which recorded, in the current mode, only light fragments were used in [225] (see Sect. 11.1.2). The experimental geometry is clear from Fig. 11.3. Of the two symmetrically placed detectors (proportional chambers), only the one corresponding to the asymmetry signs given below is shown (the plus sign indicates that light fragments are preferably emitted toward the detector). The angle θ between the neutron polarization vector and the direction at the detector D could be changed stepwise in the experiment. As a result, two independent angular distributions were added up:

$$w(\theta) = \text{const}[1 + P_n a \cos{(\theta + \theta_0)}] \ . \tag{11.3}$$

One of them corresponds to the P-odd asymmetry in the fragment emission ($\theta_0 = 0$) of the type (11.1), and the other to the P-even correlation ($\theta_0 = 90°$) of the type (10.5). The sum distribution is rotated with respect to the distribution for the purely left-right asymmetry ($\theta_0 = 90°$) by an angle arctan (a/a_{LR}), where a is the P-odd asymmetry coefficient, and a_{LR} is the P-even left-right asymmetry coefficient.

The asymmetry coefficient for the emission of light fission fragments from ^{235}U is plotted as a function of the angle θ_0 in Fig. 11.4. The left-right asymmetry coefficients for the fission of ^{233}U and ^{235}U, measured in [225], are given in Table 11.2.

Table 11.2. Left-right asymmetry coefficients a_{LR} of fission fragment emission, obtained at LINP

Compound nucleus	a_{LR}, $[10^{-4}]$	Ref.
^{234}U	-3.24 ± 0.33	[225]
	-6.43 ± 0.51	[228]
^{236}U	1.65 ± 0.11	[225]
^{240}Pu	1.25 ± 0.29	[228]
^{242}Pu	-0.36 ± 0.76	[227]

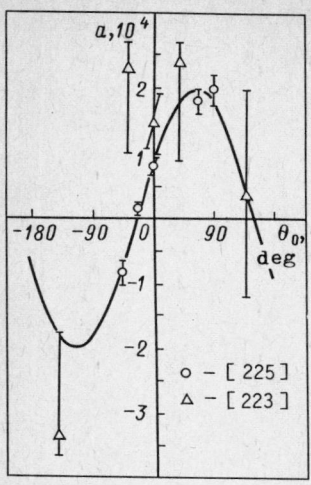

Fig. 11.4. Asymmetry coefficient in the emission of lighter fission fragments from ^{235}U as a function of the angle θ_0 [see (11.3)]. The maximum on the curve is shifted with respect to the distribution corresponding to the pure left-right asymmetry ($\theta_0 = 90°$) by an angle of $-(24.1\pm2.2)°$

The P-even asymmetry in the emission of the fission fragments of ^{233}U, ^{239}Pu, and ^{241}Pu was analyzed in [227, 228], using the techniques of mass distributions, developed for studying the asymmetry coefficient as a function of the characteristics of the output reaction channel. The results are also given in Table 11.2.

Considerable discrepancies are observed in the results for ^{234}U, but are greatly reduced if the dependence of the coefficient a_{LR} on the neutron wavelength is taken into account. Neutron wavelengths in [225] and [228] differ by a factor of 1.5.

The left-right asymmetry coefficient in the emission of ^{233}U fission fragments was studied in [231] as a function of the energy of polarized neutrons close to the resonance at $E = 0.17\,\text{eV}$. The coefficient a_{LR} increases monotonically with energy in the vicinity of the resonance.

A high left-right asymmetry in the emission of fission fragments, caused by the mixing of the levels of compound nuclei resulting from the overlapping of s- and p-wave fission resonances, indirectly confirms the hypothesis that the P-odd mixing of states occurring at the compound-nucleus stage is then "carried through" the whole fission process.

11.4 Attempt at a Theoretical Interpretation

In the preceding presentation, descriptions of experimental studies followed theoretical fundamentals (the so-called fundamentals of the method of measurements); in this chapter, we have changed this order, for several reasons. First, the violation of spatial parity in the fission of nuclei by polarized neutrons was not theoretically predicted, and the first experimental results had to be independently confirmed before they gained recognition. Second, uncertainty still exists in the interpretation of the magnitude of the experimental effects. We shall now attempt to give a phenomenological description of the process first proposed by *Sushkov* and *Flambaum* [236] (see also [126, 129]).

Sushkov and *Flambaum* [236] assumed that the fissioning nucleus transforms from the compound state, having mixed parity and an excitation energy equal to the binding energy of the neutron, to the "cold" stage characterized by a large pear-shape deformation. At this stage, the nucleus behaves like a pear-shape top; its rotational levels split at a nonzero quantum number K (projection of spin J on the revolution axis of the top) into two rotational levels of opposite parity ($\pi = \pm 1$). The wave function of the revolving nucleus can then be written in the following form [222]:

$$|aK\rangle^\pi_{JM} = (1/\sqrt{2})[|aKJM\rangle + \pi|a\overline{K}JM\rangle] , \qquad (11.4)$$

where

$$|aKJM\rangle = \sqrt{(2J+1)/4\pi}\, D^J_{MK}(\varphi,\theta,0)|aK\rangle ;$$
$$|a\overline{K}JM\rangle = |(-1)^{J+K}|a-KJM\rangle .$$

Here a is a set of parameters characterizing the internal state of the nucleus, and $D^J_{MK}(\varphi,\theta,0)$ is the rotation matrix. The levels $|aK\rangle^\pi_{JM}$ and $|aK\rangle^{-\pi}_{JM}$ are characterized by the same internal nuclear state, but differ in the macroscopic rotation (the expansion into orbital momenta includes momenta of different parities). Consequently, the amplitudes of fission of these states into any final state of the fragments coincide identically, and the wave function with the fragments removed to $r \to \infty$ has the form

$$|\widetilde{fK}\rangle_{JM} = |fK\rangle^\pi_{JM} + \beta|fK\rangle^{-\pi}_{JM} , \qquad (11.5)$$

where $\beta = \alpha(A_{-\pi}/A_\pi$, $\alpha = F\sqrt{\Delta E/D}$ is the coefficient of mixing of opposite-parity levels in a compound nucleus in the dynamic enhancement mode (see Sect. 6.6); A_π, $A_{-\pi}$ are the amplitudes of transition from the compound-nucleus state into the states $|fK\rangle^\pi_{JM}$ and $|fK\rangle^{-\pi}_{JM}$, respectively. When deriving (11.5), it was assumed that the quantum number K is conserved in the fission of the nucleus.

Squaring (11.5), we find the angular distribution of fission fragments:

$$w_{JM}(\theta) \sim |D^J_{MK}|^2(1+\gamma) + |D^J_{M,-K}|(1-\gamma) , \qquad (11.6)$$

where $\gamma = 2\,\mathrm{Re}\,\beta$.

The angular distribution of nuclear fission fragments ejected after the capture of polarized s-neutrons by nonpolarized nuclei takes the form

$$w(\theta) \approx \sum |C^{JM}_{J_i M-1/2, 1/2, 1/2}|^2 w_{JM}(\theta) \approx 1 + a_{JK}\cos\theta , \qquad (11.7)$$

where

$$a_{JK} = \gamma \frac{K}{J_i + 1/2}(-1)^{J-J_i-1/2} ,\qquad(11.8)$$

J_i is the spin of the target nucleus, and $C^{JM}_{J_i M-1/2, 1/2, 1/2}$ is the Clebsch-Gordan coefficient.

When the possible ingoing states and fission channels are taken into account, the following mean value is obtained:

$$\bar{a} = \sum_J \sum_K \omega_{JK} a_{JK} ,\qquad(11.9)$$

where ω_{JK} is the probability of fission through a channel, with given J and K.

Unfortunately, the uncertainty in the population of the spin states of compound nuclei ^{234}U, ^{236}U, and ^{240}Pu, and the lack of information on fission channels preclude unambiguous comparison of experimental data with model-based predictions. However, given certain assumptions, the formula (11.7) correctly describes the sign and relative value of the effect (see Table 11.1).

12. Spatial Parity Violation Effects in Neutron Optics

A new approach to studying P-odd effects in nuclear interactions has recently been developed: an analysis of parity violation in neutron optics. Here, the violation results from a collective action of a large number of nuclei (or atoms) in a substance; hence, the term *coherent parity violation*.

This approach dates back to *Michel*'s work in 1964 [237]. Michel showed that when a transversely-polarized neutron beam passes through a substance, the rotation of the neutron spin around the direction of the neutron momentum in the substance can be detected.

Later, *Stodolsky* [238] noticed another effect: a buildup of longitudinal polarization in an initially nonpolarized neutron beam passing through a medium. These two effects are very small in pure form, so that enhancement mechanisms had to be found.

Forte [239] observed in 1976 that these effects may be enhanced in the vicinity of the one-particle *p*-wave resonance, and proposed an experiment to measure the angle of rotation of the neutron spin in a ^{124}Sn specimen, in view of the *p*-wave resonance of ^{124}Sn at an energy of 62 eV.

The results of an experiment on neutron spin rotation, completed at ILL in 1980 [240], were unexpected. No spin rotation was observed for the ^{124}Sn nucleus, although a relatively large rotation was found in the control experiment with a target containing the natural abundance of tin isotopes. The authors replaced the ^{124}Sn specimen with a ^{117}Sn specimen, in view of the earlier-discovered large P-odd asymmetry in gamma emission from ^{117}Sn [159, 161] (see Sect. 8.2.6). The spin rotation was found to be large. The asymmetry in the transmission of longitudinally polarized neutrons of opposite helicities, predicted by *Forte* [241], was also confirmed for this nucleus.

Large effects observed for ^{117}Sn stimulated *Stodolsky* [242] to seek an explanation in terms of absolutely new weak forces. An explanation in the framework of the known interactions was found by *Sushkov* and *Flambaum* [243], who took into account the effect of the complex structure of nuclei. They were able to show that, owing to a high-level density, the mixing of states with opposite parities must be dynamically enhanced. Furthermore, they predicted that in the vicinity of the *p*-wave resonances of complex nuclei, the effects of coherent parity violation must be strongly dependent on the neutron energy and may reach 10^{-2}–10^{-1}.

The asymmetry in the transmission of longitudinally polarized neutrons through specimens with ^{117}Sn, ^{139}La, and natural-abundance mixtures of ^{79,81}Br isotopes, and an asymmetry in the cross section of the radiative capture of longitudinally polarized neutrons by the same nuclei were soon detected with reliable accuracy at LINP [244, 245].

At the same time, the asymmetry of the total cross section of the interaction between longitudinally polarized neutrons of opposite helicities and the same three nuclei was studied at JINR in the energy range of p-wave resonances [246, 247]. The asymmetry coefficients of total cross sections were very large, in the order of 10^{-3}–10^{-2}, and even reached 7×10^{-2} for ^{139}La.

12.1 Fundamentals of the Experimental Method

Coherent effects due to a P-odd weak interaction of neutrons with the target medium result from unequal refractive indices in the medium for the neutron waves with distinct spin states. The expression for the refractive index n contains the coherent scattering length for neutrons scattered by bound nuclei, $b_N = -f(0)$ [34] (see also Sect. 1.3.2):

$$n = 1 - (\lambda^2 N/2\pi)b_N = 1 + (2\pi/p^2)Nf(0) \ . \tag{12.1}$$

In (12.1), we set $\hbar = c = 1$; N denotes the number of scattering nuclei per $1\,\text{cm}^3$, p is the momentum of a neutron, and $f(0)$ is the coherent forward scattering amplitude for neutrons. Slow neutrons are known to be isotropically scattered [34], and the coherent scattering amplitude of neutrons, f, is independent of the angle. Consequently, hereafter we shall write $f(0) = f$.

Taking into account the weak interaction, we can rewrite the coherent scattering amplitude as a sum of two terms:

$$f = f_{\text{PC}} + f_{\text{PNC}} \ , \tag{12.2}$$

where f_{PC} is the parity-conserving part of the amplitude, and

$$f_{\text{PNC}} = G' \boldsymbol{sp} \tag{12.3}$$

is a small parity-nonconserving term whose sign depends on the neutron helicity. By definition, the neutron helicity is positive if $\boldsymbol{sp} > 0$, and negative if $\boldsymbol{sp} < 0$, where $\boldsymbol{s}$ is the neutron spin. The quantitiy G' is a complex constant whose magnitude must be of the order of the weak coupling constant G.

Having passed a distance l in the target, a neutron wave gains a phase $\Delta = \text{Re}(pnl)$. The phase difference for states with helicities of the opposite signs is given by

$$\Delta_+ - \Delta_- = \text{Re}[p(n_+ - n_-)l] = (2\pi/p)lN\,\text{Re}(f_+ - f_-) \ , \tag{12.4}$$

where the plus and minus signs correspond to the two helicities of the neutron.

The polarization vector of transversely polarized neutrons rotates around the momentum by an angle

$$\varphi_{PNC} = -(\Delta_+ - \Delta_-) = -\frac{2\pi}{p} lN \operatorname{Re}(f_+ - f_-) = -4\pi lN \operatorname{Re} G' . \tag{12.5}$$

At the same time, the imaginary part of the amplitude f_{PNC} is related, through the optical theorem $\sigma_t = (4\pi/p)\operatorname{Im} f(0)$, to the difference between total cross sections, $\Delta\sigma_t$, for two states of the neutron with opposite signs of helicity:

$$\Delta\sigma_t = \sigma_t^+ - \sigma_t^- = (4\pi/p)\operatorname{Im}(f_+ - f_-) = 8\pi \operatorname{Im} G' . \tag{12.6}$$

The ratio of the real to the imaginary parts of f_{PNC} follows from (12.5) and (12.6):

$$\operatorname{Re} f_{PNC}/\operatorname{Im} f_{PNC} = -(2/N)\Delta\varphi/\Delta\sigma_t , \tag{12.7}$$

where $\Delta\varphi = \varphi_{PNC}/l$ is the angle of rotation of the spin of transversely polarized neutrons per unit length of the target.

The experiments with longitudinally polarized neutrons measured three physical quantities which must be related to one another and to the spin rotation angle for transversely polarized neutrons.

The first of these is the transmission asymmetry coefficient

$$\varepsilon = (N^+ - N^-)/P_n(N^+ + N^-) , \tag{12.8}$$

where $N^\pm$ are the counting rates (or the values of beam intensity) for neutrons with opposite helicities in the beam transmitted through the target.

The second quantity is the asymmetry coefficient of total cross sections of interaction for neutrons having opposite helicities,

$$a_t = (\sigma_t^+ - \sigma_t^-)/(\sigma_t^+ + \sigma_t^-) \approx \Delta\sigma_t/2\sigma_t . \tag{12.9}$$

In usual situations, $n\Delta\sigma_t \ll 1$, where n is the target thickness in nuclei/cm^2; hence,

$$\varepsilon = -\tanh(n\Delta\sigma_t/2) \approx -n\sigma_t a_t = -a_t l/\lambda , \tag{12.10}$$

where l is the length of the target, and $\lambda = 1/n\sigma_t$ is the mean free path length of neutrons in the target.

The third physical quantity is the asymmetry coefficient of radiative capture cross sections σ_a for neutrons having opposite helicities:

$$a_a = (\sigma_a^+ - \sigma_a^-)/(\sigma_a^+ + \sigma_a^-) . \tag{12.11}$$

It is important to clarify the role of resonance effects in coherent parity violation because the high values of the observed effects (spin rotation and transmission asymmetry for polarized neutrons) can only be explained by correctly interpreting the enhancement factors due to resonance phenomena.

The possiblity of kinematic enhancement of P-parity violation effects in the vicinity of p-wave resonances was first pointed out by *Karmanov* and *Lobov* [248]. *Forte* [239, 241] reanalyzed this problem. The kinematic enhancement factor of the type $\sqrt{\Gamma_s^n/\Gamma_p^n}$ emerged when the interference of the ingoing s- and p-channels for the same exit channel was taken into account (see Sect. 6.6). Here, Γ_s^n and Γ_p^n are the neutron widths of the s- and p-resonances. At low energies, $\sqrt{\Gamma_s^n/\Gamma_p^n} \approx 1/pR$, where R is the nuclear radius, so that for nuclei with $A \approx 100$ and neutron energies of the order of 1–10 eV, this enhancement factor reaches 10–10^3.

Sushkov and *Flambaum* [126, 243] (see also [249]) remarked that in [239, 241], the effects of coherent parity violation were attributed to the interaction of neutrons with the P-odd nuclear potential; i.e., the nucleus was treated as a particle without internal degrees of freedom. The virtual excitation of the compound nucleus must give the dynamic enhancement factor $\sqrt{\Delta E/D}$ [where $\Delta E \approx 1$ MeV is the mean spacing between one-particle states, and $D \approx 1$–10 eV is the mean spacing between the levels of the compound nucleus (for medium and heavy nuclei)] of the order of 10^2–10^3 (see Sect. 6.6).

This approach to the role of the resonance effects was called the model of mixed compound states (see [250]). According to this model, the difference between the total interaction cross sections of neutrons having opposite helicities can be conveniently represented in the vicinity of p-wave resonances in the form

$$\Delta \sigma_t = 2P(E)\sigma_t(E) , \quad \text{where} \tag{12.12}$$

$$P(E) = 2\alpha \sqrt{\Gamma_s^n(E)/\Gamma_p^n(E)} \tag{12.13}$$

(α is the coefficient of level mixing in parity, see Sect. 6.6), and $\sigma_t(E)$ is the Breit-Wigner cross section close to the p-resonance. In this range $P(E)$ is practically constant and equals $P(E_p)$, where E_p is the energy of p-resonance. It is easy to see that $P(E_p)$ plays the role of the asymmetry coefficient at the resonance, $a_t(E_p)$.

We shall give the formula used to compare the asymmetry coefficients of transmission for the same nucleus but at the thermal E_{th} and resonance E_p energies:

$$a_t(E_{th})/P(E_p) = [\sigma_t(E_p)/\sigma_t(E_{th})](\Gamma_p/2E_p)^2 . \tag{12.14}$$

Finally, we give the relation between the spin rotation angle $\Delta\varphi$ and the total cross section asymmetry $\Delta\sigma_t$, in order to facilitate the comparison of the results:

$$\Delta\varphi = N\Delta\sigma_t(E - E_p)/\Gamma_p . \tag{12.15}$$

The quantity Γ_p in (12.14) and (12.15) is the total width of the p-resonance.

Stodolsky [251] analyzed the contribution of the elastic and inelastic channels of the neutron-nuclei interaction into the angle of rotation $\Delta\varphi$ of the polarization vector and into the difference of total cross sections, $\Delta\sigma_t$, and studied this contribution as a function of neutron energy. It was found that: (1) the rotation angle $\Delta\varphi$ is independent of energy in a considerable energy range, and (2) the difference between total cross sections consists of two terms: $\Delta\sigma_t = \Delta\sigma_s + \Delta\sigma_{ex}$, where σ_s is the coherent elastic scattering cross section, and σ_{ex} is the cross section of exothermal reactions, the highest of which is the radiative capture cross section σ_a; furthermore, $\Delta\sigma_s$ is proportional to the neutron momentum p, and $\Delta\sigma_{ex}$ is independent of momentum. Hence, the main contribution to $\Delta\sigma_t$ at thermal energies of neutrons must be that of the radiative neutron capture; experiments confirm this conclusion [244, 245].

A somewhat different approach to the role of resonance effects in the coherent parity violation was developed by *Zaretsky* and *Sirotkin* [252], who proposed a model of mixing of one-particle components of the compound-nucleus wave functions in the continuous spectrum in the input reaction channel. It is not yet possible, however, to discriminate among these theortical models on the basis of the experimental data available.

12.2 Survey of Experiments

12.2.1 Neutron Spin Rotation

The first experiment on the observation of polarization vector rotation in a transversally polarized neutron beam was completed in 1980 at ILL [240]. The unexpected discovery of this effect in a ^{117}Sn target was described at the beginning of this chapter. Later, a second experiment was run at ILL with a modified apparatus, and the spin rotation was observed in targets containing natural isotopic abundances of lead and tin, respectively [253]. We shall describe this experiment in some detail.

The experiment used a cold polarized neutron beam (entry 14 in Table 1.1). The apparatus (Fig. 12.1) was a neutron polarimeter; i.e., it used a crossed polarizer and analyzer. A neutron selector installed at the beam entrance formed a quasimonochromatic neutron beam with wavelengths greater than the Bragg threshold neutron wavelength in lead and tin (about 0.5 nm; see [34]). Neutron beam monitors were placed in front of and behind the selector. The beam was then polarized by a system of vertically arranged supermirrors (see Sect. 1.3.5), magnetized along the z-axis. Polarized neutrons entered the region of low magnetic field ($H < 4 \times 10^{-4}$ A/cm) formed within a triple-layer mumetal shield.

The system of a crossed polarizer and analyzer was formed by two identical coils with current sheets at the end faces, forming the regions through which neutrons passed nonadiabatically. The input coil kept the neutron spins pointing along z, and the output coil was rotated by $90°$ relative to the input one, so that its field pointed along the y-axis and reversed the direction ($\pm y$)

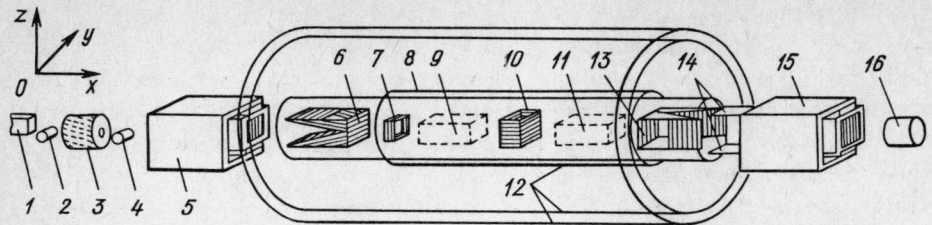

Fig. 12.1. The ILL experiment measuring the neutron spin rotation angle [253]: *(1)* neutron guide, *(2)* first monitor, *(3)* neutron velocity selector, *(4)* second monitor, *(5)* polarizing supermirrors, *(6)* input current-carrying coil, *(7)* trim coil, *(8)* region of low magnetic field, *(9)* first position of the target, *(10)* π-coil, *(11)* second position of the target, *(12)* three-layer mumetal shield, *(13)* output current-carrying coil, *(14)* shims for rotating the field, *(15)* analyzing supermirrors, *(16)* neutron counter

when the current was reversed. The magnetic field was then gradually rotated back toward the z-axis by thin iron shim plates, so that it pointed along the direction of the magnetic field in the analyzing mirrors.

The output coil together with the analyzing mirrors and the neutron counter thus formed a device for detecting the rotation of neutron spins in the target around the beam direction. A reversal of current in the output coil resulted in reversing the direction of the analyzed polarization component. The degree of spin rotation in the target was deduced from the difference between counting rates for two current directions in the output coil.

An auxiliary π-coil was introduced to eliminate instrument effects due both to the residual magnetic field inside the shield and the nonperpendicularity of the input and output coils. The magnetic field of the auxiliary coil was vertical (parallel to the z-axis), and the field strength was chosen so as to have the Larmor precession of neutron spins rotate their projection on the plane xy around the z-axis exactly through 180° over the time of flight through the coil.

Two identical targets were alternately inserted into the beam, in front of and behind the π-coil. Therefore, when the target rotated the spin in the plane yz, the spin projection on the y-axis changed sign when the target position was changed, but preserved it when the spin was rotated in the residual magnetic field by an angle independent of the target position. In the former case, the spin rotation effects added up; in the latter case, they cancelled out in the subtraction of the results obtained for the two positions of the target.

The angle of rotation φ was found from the formula

$$\sin \varphi = (N_+ - N_-)/P_{n1}P_{n2}(N_+ + N_-) \;, \tag{12.16}$$

where $\sin \varphi$ is the projection of the neutron spin unit vector on the y-axis in the region of the output coil, $N_+(N_-)$ is the neutron counting rate for the positive (negative) current direction in the output coil, and $P_{n1}P_{n2}$ are the polarization efficiencies of the polarizing and analyzing supermirrors (see Sect. 1.1).

The main control experiment consisted in accumulating the data with the π-coil switched off. The spin rotation effects were then identical for the two target positions, and cancelled out when the corresponding results were subtracted.

The output coil current was reversed once a second. Targets were alternated once a minute. Every five days, the tin (lead) specimens at the first and second positions were switched and their orientation in space was altered (the entrance face was turned to become the exit face). Two targets with natural abundances of tin isotopes and six targets with natural abundances of led isotopes were used, having different (but pairwise identical) length and structure.

The results of measurements are given in Table 12.1 (see below).

Table 12.1. Results of measurements of neutron spin rotation angle $\Delta\varphi$ at ILL

Target nucleus	$\Delta\varphi$, $[10^{-6}\,\text{rad/cm}]$	Ref.
^{124}Sn	-0.48 ± 1.49	[240]
^{117}Sn	-37.0 ± 2.5	[240]
Sn$_{\text{nat}}$	-3.19 ± 0.40	[253]
Pb$_{\text{nat}}$	2.24 ± 0.33	[253]
^{139}La	-219 ± 29	[255]

12.2.2 Transmission of Longitudinally Polarized Neutrons

The asymmetry in the transmission of neutrons having opposite helicities was observed in [240]. The apparatus was modified to produce a longitudinally polarized beam incident on the target. This was achieved by adding a trim coil which turned neutron spins along the y-axis (see Fig. 12.1), and by reducing the current in the π-coil so as to have the neutron spins turn in the direction of the x-axis. The reversal of current, once a second, in the π-coil reversed the helicity of the neutrons. A ^{117}Sn specimen was invariably in the second position. The output coil and the analyzing mirrors were removed. The control experiment was run with a depolarized beam. The asymmetry coefficient for the transmission of opposite-helicity neutrons, ε, was measured [see (12.8)]. The result obtained for the polarized beam was $\varepsilon = -(9.78\pm4.01)\times10^{-6}$, and, for the depolarized beam, $\varepsilon = -(2.99\pm5.33)\times10^{-6}$.

A more accurate value of the asymmetry coefficient for the total cross section of interaction of opposite-helicity neutrons, a_t, was obtained at LINP [244, 245], where the asymmetry coefficient for the radiative-capture cross section, a_a, was simultaneously measured. Measurements were carried out with ^{117}Sn, ^{139}La, and with a natural-abundance mixture of ^{79}Br and ^{81}Br.

The LINP apparatus is shown in Fig. 12.2. A polarized neutron beam, of 6×10^7 neutrons/s flux and 6×1 cm cross section, passed through an adiabatic high-frequency spin flipper (see Sect. 1.2.3). When the hf field was switched on,

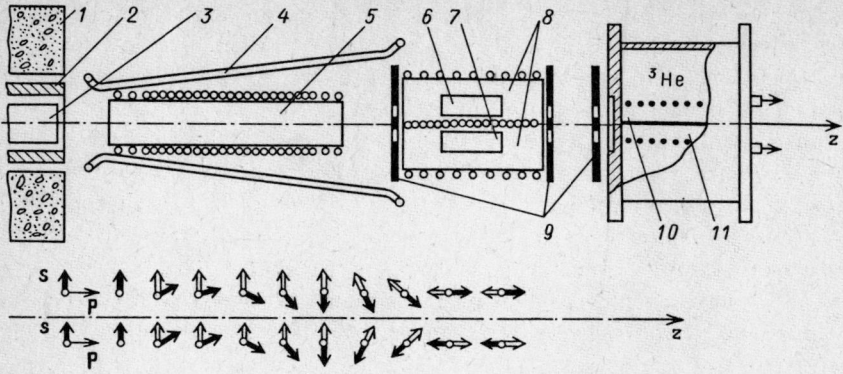

Fig. 12.2. The LINP experiment measuring the transmission asymmetry for neutrons having opposite helicities [244]. The evolution of the orientation of the neutron spin *s* passing through the adiabatic spin flipper is shown without rf field (white arrows) and with rf field (black arrows) for specimens 6 and 7, respectively: *(1)* shielding, *(2)* permanent magnet, *(3)* neutron guide, *(4)* spin flipper coils, *(5)* rf coil of the spin flipper, *(6,7)* specimens, *(8)* solenoids producing longitudinal magnetic fields with opposite directions, *(9)* diaphragms made of ^{6}LiF, *(10,11)* neutron detectors

the spin flipper rotated the neutron spins against the direction of the magnetic field; otherwise, the spins followed the field.

The beam was then split into two beams by lithium diaphragms; the two beams were incident on two identical specimens located within two solenoids which produced longitudinal magnetic fields in the opposite directions. Therefore, two identical specimens were irradiated by two neutron beams having opposite helicities. Two multiwire proportional chambers filled with a mixture of ^{3}He, CO_2, and Ar were used as neutron detectors.

Current signals from the chambers were fed to a differential amplifier in which the currents were subtracted. The signals from each chamber and the difference signal were integrated over 2 s, converted into digital code, and stored on magnetic tape for subsequent analysis.

The separation into two channels served to eliminate the effects of beam intensity fluctuations. The sensitivity of the system was increased by a factor of 3–5, and it was now possible to suppress spurious effects.

Every 12 hours, the direction of current in both solenoids was reversed, constituting a control experiment: the sign of the asymmetry effect was thereby reversed. A control experiment was also run with a depolarized beam.

Scintillation detectors of gamma quanta with NaI(Tl) crystals were installed on both sides of one of the specimens, in order to analyze the asymmetry in the radiative capture cross section for neutrons with opposite helicities. The current signals from photomultipliers were fed to a differential amplifier; a signal proportional to the reactor power, and hence, to the neutron beam intensity, was fed to the other input of the amplifier. The results of the LINP experiments on the asymmetry of total-interaction cross sections and of radiative-capture cross sections are given in Table 12.2.

Table 12.2. Total-interaction cross section asymmetry coefficient a_t and radiative-capture cross section asymmetry coefficient a_a for thermal neutrons having opposite helicities

Target nucleus	$\sigma_t(E_{th})$ [10^{-24} cm^2]	a_t, [10^{-6}]	a_a, [10^{-6}]	Source
^{79,81}Br	15.5	9.8±1.0	15.5±1.5	LINP [245]
	12.8	9.5±1.7		ITEP [254]
^{117}Sn	3.7	6.2±0.7	22.6±1.9	LINP [244]
	4	11.2±2.6		ITEP [254]
^{139}La	19	9.0±1.4	16.1±2.0	LINP [244]

At JINR [246, 247], the transmission of longitudinally polarized neutrons of opposite helicities was measured directly in the p-resonances of the same nuclei ^{117}Sn, ^{139}La, ^{81}Br, and also ^{111}Cd. The experiment was carried out at the IBR-30 pulsed reactor. The resonance neutrons were polarized by passing the beam through a polarized proton target. The characteristics of this beam are shown in Table 1.1 (see entry 10). The longitudinal polarization of neutrons and the flipping of the polarization vector were produced by the appropriate configurations of the leading magnetic field. Neutrons with a specified energy were selected by the time-of-flight technique, over a flight base, 58 m long. Polarization was flipped every 40 s. In order to reverse the sign of the transmission asymmetry effect without changing the equipment and the geometry of the experiment, the proton target polarization was reversed every two days.

Figure 12.3 plots the energy dependence of the asymmetry coefficient ε for the transmission of neutrons, having opposite helicities, in the region of the p-wave resonance of the ^{139}La nucleus, and also a part of the spectrum of detector counts with respect to the time of flight in the vicinity of the p-resonance. If the resonances are not broadened by the Doppler effect and by an insufficient

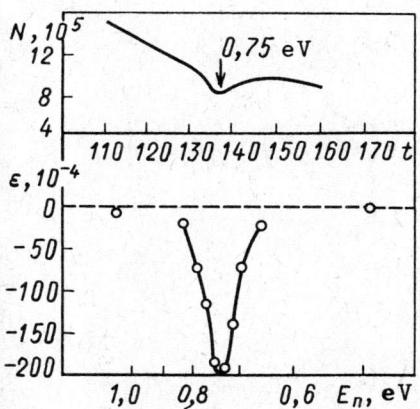

Fig. 12.3. Parity nonconservation effect as a function of energy in the range of the p-wave resonance of ^{139}La. The upper curve gives the neutron spectrum vs. time of flight t; the lower curve plots the transmission asymmetry coefficient ε

energy resolution, the transmission is related to $P(E)$ by (12.10) and (12.12). Otherwise, the expression is considerably more complex [247].

Furthermore, *Alfimenkov* et al. [246, 247] also measured the parameters of p-resonances required to compare the experimental data in the thermal and resonance energy regions, both with each other and with the theoretical predictions. The results of the JINR experiments are shown in Table 12.2.

The asymmetry of total cross sections for the interaction of neutrons having opposite helicities with a number of nuclei was studied at ITEP [254]. For the ^{232}Th and ^{239}Pu nuclei and for the natural-abundance mixture of chlorine and lead isotopes, the upper bounds on the asymmetry coefficient were found to be $(2-7) \times 10^{-6}$. The results obtained for the ^{117}Sn nucleus are shown in Table 12.2.

12.3 Summary of Experimental Data

The results of the described experiments are summarized in Tables 12.1–3.

The positive sign corresponds to the right-handed rotation of spin in the target around the momentum of the neutron. The large value of $\Delta\varphi$ for the natural-abundance mixture of lead isotopes has not yet been explained, because no p-wave resonances at low-neutron energy were observed for any lead isotope. The data for the natural-abundance mixture of tin isotopes are explained in terms of the results obtained for ^{117}Sn and by the abundance of ^{117}Sn (7.6 %) in the natural mixture of tin isotopes.

A comparison of results for a_t and a_a shows that at least 70 % of the effects in total-interaction cross sections can be attributed to the effects in radiative-capture cross sections.

The prediction and explanation of such large effects in p-resonances is given, as described in Sect. 12.1, in the model of mixed compound states when both kinematic and dynamic enhancement factors are taken into account [126, 243].

Using these results, *Alfimenkov* et al. estimated in [247] and in the review paper [250] the matrix elements of the weak interaction which produces the mixing of the compound s- and p-states $\langle s|H_w|p\rangle$ (see Sect. 6.6). It was men-

Table 12.3. Results of measurements of total-interaction cross section asymmetry coefficient $P(E)$ for resonance neutrons having opposite helicities, and of the parameters of the p-resonances at JINR [246, 247]

Target nucleus	E_p, [eV]	$\sigma_t(E_p)$ [10^{-24} cm^2]	Γ_p, [10^{-3} eV]	$P(E_p)$, [10^{-3}]
^{81}Br	0.88±0.01	0.9±0.1	190±20	24 ± 4
^{111}Cd	4.53±0.03	3.8±0.5	160±10	-8.6 ± 1.2
^{117}Sn	1.33±0.01	1.6±0.2	230±20	4.5 ± 1.3
^{139}La	0.75±0.01	2.8±0.4	45±5	73 ± 5

Table 12.4. Comparison of total-interaction cross section asymmetry coefficients measured for neutrons having opposite helicities, in the thermal and resonance energy ranges

Target nucleus	a_t, $[10^{-6}]$	Recalculated to $a_t^*(E_{th})$, $[10^{-6}]$
^{79,81}Br	9.8±1.0 [245] 9.5±1.7 [254]	9.8±2.9^a [247]
^{117}Sn	6.2±0.7 [244] 11.2±2.6 [254]	14.5±5.5 [247] 13.5±3.9 [247, 254]
^{139}La	9.0±1.4 [244]	9.3±2.9 [247]

^{a} The result is obtained by recalculating for the natural-abundance mixture of bromine isotopes, ^{79,81}Br.

tioned, however, that this calculation was not unambiguous because both the spins of the p-resonances and some other required parameters remain unknown.

It is interesting to compare P-odd effects obtained at different energies on the same nucleus. This comparison is shown in Table 12.4, where we give the asymmetry coefficient $a_t^*(E_{th})$ calculated on the basis of the measured asymmetry coefficient $P(E_p)$ in the resonance region of energies, using the formula (12.14).

Clearly, the results are in good agreement for ^{79,81}Br and ^{139}La. The results for ^{117}Sn [247, 254] agree with the results of calculations via (12.14).

Let us compare the results of measurements of spin rotation of transversally polarized neutrons with the calculated results based on the measured asymmetry in the total cross sections of the interaction between neutrons having opposite helicities for ^{117}Sn nuclei [250]. This comparison is given below, where the $\Delta\varphi^*$ column gives rotation angles calculated via (12.15):

$\Delta\varphi [10^{-6}\,\text{rad/cm}]$	$\Delta\varphi^* [10^{-6}\,\text{rad/cm}]$
	−23 ±10 [253] −11 ± 2 [244]
−37.0±2.5 [253]	
	−25 ±11 [246] −18.8 ± 4.4 [254]

The results are in satisfactory agreement; the one obtained by recalculating result [244] is to some extent an exception.

We conclude that a comparison of all available experimental results shows that the model of mixed compound states [126, 243] adequately describes the energy dependence of the effect, provided that we take into account that the magnitudes of the effects in the resonance and thermal energy ranges differ by 3 to 4 orders.

Recently, *Alfimenkov* et al. [256] at JINR detected a left-right asymmetry in the emission of $E_\gamma = 9.32$ gamma quanta in the radiative capture of polarized neutrons in the neutron p-wave resonance of ^{117}Sn at $E_p = 1.33\,\mathrm{eV}$. The magnitude of this asymmetry made it possible to calculate the fraction of the partial neutron width $\Gamma^n_{p\,1/2}$ of this resonance, which goes through the channel with the total neutron momentum $j = 1/2$, with respect to the total neutron width Γ^n_p i.e. $\Gamma^n_{p\,1/2}/\Gamma^n_p = 0.27\pm0.03$.

We should note in conclusion that the study of the violation of spatial parity in neutron optics opens up new possibilities for the analysis of the structure and excited states of nuclei. Suffice it to recall that the existence of low-energy p-wave resonances in the ^{81}Br, ^{111}Cd, ^{117}Sn, and ^{139}La nuclei was discovered in the course of the study of P-odd effects. We can also expect that the accumulation of experimental data in this field may reveal new information on the structure of the nucleon-nucleon weak interaction.

Note added in proof: The coefficient a_t of the total cross section has recently been measured for opposite helicity thermal neutrons [Yu.A. Mostovoi, O.V. Khakhan: Yad. Fiz. **43**, 3 (1986)]. The reported results are: $a_t = (11.2\pm0.9)\cdot 10^{-6}$ for ^{79,81}Br (natural-abundance mixture), $(6.9\pm0.8)\cdot 10^{-6}$ for ^{117}Sn, and $(0.4\pm0.5)\cdot 10^{-6}$ for ^{207}Pb.

References

1. V.B. Berestetsky, E.M. Lifshitz, L.P. Pitaevsky: *Relativistic Quantum Theory*, Vol. 1 (Pergamon, Oxford, New York 1971)
2. R.R. Newton, C. Kittel: Phys. Rev. **74**, 1604 (1948)
3. E. Majorana: Nuovo Cimento **9**, 43 (1932)
4. Yu.G. Abov, A.D. Gulko, P.A. Krupchitsky: *Polarized Slow Neutrons* (Atomizdat, Moscow 1966) (in Russian)
5. W.B. Dress, P.D. Miller, I.H. Penelbury, P. Perrin, N.F. Ramsey: Phys. Rev. **D15**, 9 (1977)
6. V.M. Lobashev: Yad. Fiz. **2**, 957 (1965)
7. Yu.G. Abov, P.A. Krupchitsky, M.I. Bulgakov, O.N. Ermakov, I.L. Karpikhin: Phys. Lett. **B27**, 16 (1968); and Yad. Fiz. **10**, 558 (1969)
8. A.D. Gulko, O.N. Ermakov, I.L. Karpikhin, P.A. Krupchitsky, Yu.E. Kuznetsov, V.F. Perepelitsa: Preprint ITEP 46 (Moscow 1981)
9. H. Marshak, H. Postma, V.L. Saylor, F.J. Shore, C.A. Reynolds: Phys. Rev. **128**, 1287 (1962)
10. V.P. Alfimenkov, V.I. Lushchikov, V.G. Nikolenko, Yu.V. Taran, F.L. Shapiro: Yad. Fiz. **3**, 55 (1966)
11. L. Alvarez, F. Bloch: Phys. Rev. **57**, 111 (1940)
12. W.R. Arnold, A. Roberts: Phys. Rev. **71**, 878 (1947)
13. F. Bloch, D.B. Nicodemus, H.H. Staub: Phys. Rev. **74**, 1025 (1948)
14. V.V. Vladimirsky: Zh. Eksp. Teor. Fiz. **39**, 1062 (1960)
15. J.W.T. Dabbs, L. Roberts, S. Bernstein: Rep. ORNL-CF-1955-5-126 (1955)
16. A.D. Gulko, S.S. Trostin, A. Khudoklin: Nucl. Instrum. Methods **34**, 88 (1965)
17. M.I. Bulgakov, A.D. Gulko. S.S. Trostin: Preprint ITEP 100 (Moscow 1981)
18. K. Abrahams, O. Steinsvoll, P.J.M. Bongaarts, P.W. Lange: Rev. Sci. Instrum. **33**, 524 (1962)
19. Yu.G. Abov, M.M. Danilov, O.N. Ermakov, I.L. Karpikhin, V.K. Rissukhin, A.M. Skornyakov: Yad. Fiz. **16**, 1218 (1972)
20. G.M. Drabkin, E.I. Zabidarov, Ya.A. Kasman, A.I. Okorokov: Zh. Eksp. Teor. Fiz. **56**, 478 (1969)
21. P. Liaud, R.I. Steinberg, B. Vignon: Nucl. Instrum. Methods **125**, 7 (1975)
22. D.A. Korneev: Nucl. Instrum. Methods **169**, 65 (1980); and D.A. Korneev, V.A. Kudrjashev: Nucl. Instrum. Methods **179**, 509 (1981)
23. F. Mezei: Z. Phys. **255**, 146 (1972)
24. G. Badurek, G.P. Westphal, P. Ziegler: Nucl. Instr. Methods **120**, 351 (1974)
25. B.Van Laar, F. Maniawski, P. Mijnarends: Nucl. Instrum. Methods **133**, 241 (1976)
26. A.I. Egorov, V.M. Lobashev, V.A. Nazarenko, G.D. Porsev, A.P. Serebrov: Yad. Fiz. **19**, 300 (1974)
27. A. Abragam: *Principles of Nuclear Magnetism* (Oxford Univ. Press, London 1961)
28. Yu.V. Taran: Preprint JINR P3-8577 (Dubna 1975)
29. G.E. Bacon: *Neutron Diffraction*, 3rd edn. (Clarendon, Oxford 1975); and J.B. Hayter: in *Topics in Current Physics*, Vol. 6, ed. by H. Dachs (Springer, Berlin, Heidelberg, New York 1978); and Yu.A. Izumov, V.E. Naish, R.P. Ozerov: *Neutrons and Solids*, Vol. 2, *Neutronography of Magnetics* (Atomizdat, Moscow 1981) (in Russian)
30. F.J. Webb: "Reactor science and technology." J. Nucl. Energy, Pts. A/B **17**, 187 (1963)

31 W. Mampe, P. Ageron: Inst. Phys. Conf. Ser. Inst. Phys. London 42, 148–156 (1978)
32 J. Christ, T. Springer: Nucleonik 4, 23 (1962)
33 H. Meier-Leibnitz, T. Springer: "Reactor science and technology." J. Nucl. Energy, Pts A/B 17, 217 (1963)
34 I.I. Gurevich, L.V. Tarasov: *Low-Energy Neutron Physics* (North-Holland, Amsterdam 1968)
35 H.B. Möller, L. Passel, F. Stecher-Rasmussen: "Reactor science and technology." J. Nucl. Energy, Pts A/B 17, 227–231 (1963);
and K. Abrahams, W. Ratynski, F. Stecher-Rasmussen, E. Warming: Nucl. Instrum. Methods 45, 293 (1966)
36 F. Stecher-Rasmussen, K. Abrahams, J. Kopecky: Nucl. Phys. A181, 225 (1972)
37 D.J. Hughes, M.T. Burgy: Phys. Rev. 76, 1413 (1949); and Phys. Rev. 81, 498 (1951)
38 K. Berndorfer: Z. Phys. 243, 188 (1971)
39 M. Hetzelt, A. Heidemann: Nucl. Instrum. Methods 133, 51 (1976)
40 B.G. Erozolimsky, Yu.A. Mostovoi, B.A. Obinyakov. S.A. Petushkov, V.P. Fedunin, O.V. Khakhan, V.N. Chernishevich: Prib. Tekh. Eksp. 6, 39 (1976)
41 G.M. Drabkin, A.I. Okorokov, A.F. Shchebetov, N.V. Borovikova, A.G. Gusakov, V.A. Kudrayashev, V.V. Runov, D.A. Korneev: Nucl. Instrum. Methods 133, 453 (1976); Zh. Tekh. Fiz. 47, 203 (1977)
42 A.P. Bulkin, V.Ya. Kezerashvili, V.A. Kudryavtsev, A.N. Pirozhkov, V.G. Siromyatnikov, V.P. Kharchenkov, A.F. Shchebetov: Nucl. Instrum. Methods 178, 105 (1980)
43 B. Hamelin: Nucl. Instrum. Methods 135, 299 (1976)
44 J.B. Hayter, J. Penfold, W.G. Williams: J. Phys. E: Sci. Instrum. 11, 454 (1978)
45 B.P. Schoenborn, D.L. Caspar, O.F. Kammerer: J. Appl. Crystallogr. 7, 508 (1974)
46 J.W. Lynn, J.K. Kjems, L. Passell: J. Appl. Crystallogr. 9, 454 (1976)
47 A.G. Gusakov, V.V. Deriglasov, V.Ya. Kezerashvili, G.A. Krutov, V.A. Kudryavtsev, B.P. Peskov, V.G. Siromyatnikov, V.A. Trunov, V.P. Kharchenkov, A.F. Shchebetov: Zh. Eksp. Teor. Fiz. 77, 1720 (1979)
48 V.F. Turchin: At. Energ. 22, 119 (1967)
49 F. Mezei: Commun. Phys. 1, 81 (1976)
50 F. Mezei, P.A. Dagleish: Commun. Phys. 2, 41 (1977)
51 F. Mezei: Inst. Phys. Conf. Ser. Inst. Phys. London 42, 162–168 (1978)
52 Yu.G. Abov, M.I. Bulgakov, A.D. Gulko, O.N. Ermakov, P.A. Krupchitsky, Yu.A. Oratovsky, S.S. Trostin: Prib. Tekh. Eksp. 4, 195 (1966)
53 C. Christensen, V. Krohn, R. Ringo: Phys. Lett: B28, 411 (1969); and Phys. Rev. C1, 1693 (1970)
54 V.P. Alfimenkov, S.B. Borzakov, Ya. Werzhbitski, A.I. Ivanenko, Yu.D. Mareev, N.I. Moreva, O.N. Ovchinnikov, L.B. Pikelner, E.I. Sharapov: Preprint JINR P3-12040 (Dubna 1978)
55 *Annual Report 1978* (ILL, Grenoble, France 1978) pp. 28–29
56 *Annual Report 1981* (ILL, Grenoble, France 1981) pp. 33, 43
57 K. Dörr, H. Ackermann, B. Bader, H.-J. Stöckmann, P.V. Blanckenhagen: Nucl. Instrum. Methods 190, 211 (1981)
58 L.B. Okun: *Leptons and Quarks* (North-Holland, Amsterdam 1982)
59 O.R. Frisch, O. Stern: Z. Phys. 85, 4 (1933)
60 J.R. Dunninig, P.N. Powers, H.G. Beyer: Phys. Rev. 51, 51 (1937);
and J.G. Hoffman, M.S. Livingston, H.A. Bethe: Phys. Rev. 51, 214 (1937);
and P.N. Powers, H.G. Beyer, J.R. Dunning: Phys. Rev. 51, 371 (1937)
61 O.R. Frisch, H. Halban, J. Koch: Nature (London) 139, 756, 1021 (1937); and Nature (London) 140, 360 (1937); and Phys.Rev. 53, 719 (1938)
62 I.I. Rabi: Phys. Rev. 51, 652 (1937); and
I.I. Rabi, S. Millman, P. Kusch, J.R. Zacharias: Phys. Rev. 55, 526 (1939)
63 E.M. Purcell, N.F. Ramsey: Phys. Rev. 78, 807 (1950)
64 L.D. Landau: Zh. Eksp. Teor. Fiz. 32, 405 (1957); and Nucl. Phys. 3, 137 (1957)
65 J.H. Christenson, J.W. Cronin, V.L. Fitch, R. Turlay: Phys. Rev. Lett. 13, 138 (1964)
66 H. Kopfermann: *Kernmomente* (Akademische Verlagsgesellschaft, Frankfurt/M. 1956)
67 F. Bloch: Phys. Rev. 70, 460 (1946)
68 N.F. Ramsey: Phys. Rev. 78, 695 (1950)
69 N.F. Ramsey: *Molecular Beams* (Clarendon Press, Oxford 1956)
70 V.W. Cohen, V.R. Corngold, N.F. Ramsey: Phys. Rev. 104, 283 (1956)

71 G.L. Greene, N.F. Ramsey, W. Mampe, J.M. Pendlebury, K. Smith, W.B. Dress, P.D. Miller, P. Perrin: Phys. Rev. **D20**, 2139 (1979)
72 P.J. Winkler, D. Kleppner, T. Myint, F.G. Walther: Phys. Rev. **A5**, 83 (1972)
73 E.H. Rogers, H.H. Staub: Phys. Rev. **76**, 980 (1949)
74 D.A. Korneev, A.V. Petrenko: Preprint JINR P3-82-102 (Dubna 1983)
75 Yu.A. Aleksandrov: *The Fundamental Properties of the Neutron* (Energoizdat, Moscow 1982) (in Russian)
76 J.H. Smith, E.M. Purcell, N.F. Ramsey: Phys. Rev. **108**, 120 (1957)
77 W.B. Dress, J.K. Baird, P.D. Miller, N.F. Ramsey: Phys. Rev. **170**, 1200 (1968); and Phys. Rev. **179**, 1285 (1969)
78 C.G. Shull, R. Nathans: Phys. Rev. Lett. **19**, 384 (1967)
79 F.L. Shapiro: Usp. Fiz. Nauk **95**, 146 (1968)
80 I.S. Altarev, Yu.V. Borisov, N.V. Borovikova, A.B. Brandin, A.I. Egorov, V.F. Ezhov, V.M. Lobashev, V.A. Nazarenko, V.L. Ryabov, A.P. Serebrov, R.R. Taldaev: Phys. Lett. **B102**, 13 (1981)
81 S. Weinberg: Phys. Rev. Lett. **37**, 657 (1976)
82 S.L. Glashow: Nucl. Phys. **22**, 579 (1961);
and S. Weinberg: Phys. Rev. Lett. **19**, 1264 (1967);
and A. Salam: in *Elementary Particle Theory*, ed. by N. Svartholm (Almquist & Wiksell, Stockholm 1968) pp. 367–378
83 H.A. Tolhoek, J.A.M. Cox: Physica **19**, 101 (1953)
84 M.E. Rose: Phys. Rev. **75**, 213 (1949)
85 S. Bernstein, L.D. Roberts, C.P. Stanford, J.W.T. Dabbs, T.E. Stephenson: Phys. Rev. **94**, 1243 (1954)
86 L.D. Roberts, S. Bernstein, J.W.T. Dabbs, C.P. Stanford: Phys. Rev. **95**, 105 (1954)
87 J.W.T. Dabbs. L.D. Roberts, S. Bernstein: Phys. Rev. **98**, 1512 (1955)
88 A. Stolovy: Phys. Rev. **118**, 211 (1960);
and H. Marshak, H. Postma, V.L. Saylor, F.J. Shore, C.A. Reynolds: Phys. Rev. **128**, 1287 (1962); and Phys.Rev. **126**, 979 (1962);
and V.L. Saylor, R.L. Schermer, F.J. Shore, C.A. Reynolds, H. Marshak, H. Postma: Phys. Rev. **127**, 1124 (1962);
and G. Brunhart, H. Postma, V.L. Saylor: Phys. Rev. **B137**, 1484 (1965)
89 Yu.V. Taran, F.L.Shapiro: Zh. Eksp. Teor. Fiz. **44**, 2185 (1963)
90 P. Draghicescu, V.I. Lushchikov, V.G. Nikolenko, Yu.V. Taran, F.L. Shapiro: Phys. Lett. **12**, 334 (1964)
91 V.I. Lushchikov, Yu.V. Taran, F.L. Shapiro: Yad. Fiz. **10**, 1178 (1969)
92 A. Abragam, M. Borghini: in *Progress in Low Temperature Physics*, Vol. 4, ed. by C.J. Gorter, (North-Holland, Amsterdam 1964) pp. 384–449
93 V.P. Alfimenkov, V.I. Lushchikov, V.G. Nikolenko, Yu.V. Taran, F.L. Shapiro: Phys. Lett. **B24**, 151 (1967)
94 V.P. Alfimenkov, G.G. Akopyan, V.A. Vagov, A.I. Ivanenko, L. Lason, Yu.D. Mareev, N.I. Moreva, O.N. Ovchinnikov, L.B. Pikelner, S. Salai, E.I. Sharapov: Yad. Fiz. **25**, 930 (1977);
and G.G. Akopyan, V.P. Alfimenkov, Ya. Werzhbitski, A.I. Ivanenko, Yu.D. Mareev, N.I. Moreva, O.N. Ovchinnikov, L.B. Pikelner, E.I. Sharapov: Yad. Fiz. **26**, 942 (1977)
95 V.P. Alfimenkov, L.B. Pikelner,E.I. Sharapov: Fiz. Elem. Chastits At. Yadra **11**, 2, 411 (1980)
96 V.P. Alfimenkov, O.N. Ovchinnikov: Preprint JINR P8-9168 (Dubna 1975);
and B.S. Neganov, N. Borisov, M. Linburg: Zh. Eksp. Teor. Fiz. **50**, 1445 (1966)
97 G.A. Keyworth, C.E. Olsen, F.T. Seibel, J.W.T. Dabbs, N.W. Hill: Phys. Rev. Lett. **31**, 1077 (1973);
and G.A. Keyworth, J.R. Lembey, C.E. Olsen, F.T. Seibel, J.W.T. Dabbs, N.W. Hill: Phys. Rev. **C18**, 1328 (1978)
98 E.R. Reddingius, H. Postma, C.E. Olsen, D.C. Rorer, V.L. Saylor: Nucl. Phys. **A218**, 84 (1974)
99 V.G. Baryshevsky, M.I. Podgoretsky: Zh. Eksp. Teor. Fiz. **47**, 1050 (1964)
100 A. Abragam, G.L. Bacchella, H. Glättli, P. Meriel, J. Piesvaux, M. Pinot: J. Phys. Lett. **36**, 263 (1975);
and H. Glättli, A. Abragam, G.L. Bacchella, M. Fourmond, P. Meriel, J. Piesvaux,

M. Pinot: Phys. Rev. Lett. **40**, 748 (1978);
and H. Glättli: 2nd Int. Neutron Physics School (Coll. Lect.) Dubna, 1974, pp. 403–419;
and H. Glättli, G.L. Bacchella, M. Fourmond, A. Malinovski,P. Meriel, M. Pinot, P. Roubeau, A. Abragam: J. Phys. **40**, 629 (1979);
and H. Glättli, J. Coustham: J. Phys. **44**, 957 (1983)

101 H.J. Ligthart, H. Postma: Z. Phys. **A288**, 179 (1978)
102 A.J. Ferguson: *Angular Correlation Methods in Gamma-Ray Spectroscopy* (North-Holland, Amsterdam 1965)
103 E.R. Reddingius, J.J. Bosman, H. Postma: Nucl. Phys. **A206**, 145 (1973)
104 J. Honzatko, J. Šebek, J. Kajfosz, J. Stehno, Z. Kosina, K. Konecny: Nucl. Phys. **A209**, 245 (1973)
105 J.J. Bosman, H. Postma: Nucl. Instrum. Methods **148**, 331 (1978)
106 P.P.J. Delheij, A. Girgin, K. Abrahams, H. Postma, W.J. Huiskamp: Nucl. Phys. **A341**, 21 (1980);
and P.P.J. Delheij, K. Abrahams, W.J. Huiskamp, H. Postma: Nucl. Phys. **A341**, 45 (1980)
107 H. Schopper: Nucl. Instrum. **3**, 158 (1958)
108 G. Trumpy: Nature (London) **176**, 507 (1955)
109 G. Trumpy: Nucl. Phys. **2**, 664 (1956/1957)
110 R. Michalec, T. Ruskov: Czech. J. Phys. **12**, 325 (1962)
111 J. Vervier: Nucl. Phys. **26**, 10 (1961)
112 J. Kopecky, J. Kajfosz, B. Chalupa: Nucl. Phys. **68**, 449 (1965)
113 K. Abrahams, W.Ratynski:Nucl. Phys. **A124**, 34 (1969);
and F. Stecher-Rasmussen, K. Abrahams, J. Kopecky: Nucl. Phys. **A181**, 225–240, 241–249, 250–261 (1969); and Nucl. Phys. **A188**, 535 (1972); and Nucl. Phys. **A215**, 45 (1973);
and A.M.J. Spits, J. DeBoer: Nucl. Phys. **A224**, 517 (1974);
and A.M.J. Spits, J. Kopecky: Nucl. Phys. **A264**, 63 (1976);
and J. DeBoer, K. Abrahams, J. Kopecky, P.M. Endt: Nucl. Phys. **A352**, 125 (1981)
114 R. Vennink, W. Ratynski, J. Kopecky: Nucl. Phys. **A299**, 429 (1978)
115 V.A. Vesna, E.A. Kolomensky, V.B. Kopeliovich, V.M. Lobashev, V.A. Nazarenko, A.N. Pirozhkov, E.V. Shulgina: Nucl. Phys. **A352**, 181 (1981)
116 C.S. Wu, E. Ambler, R.W. Hayward, D.D. Hoppes, R.P. Hudson: Phys. Rev. **105**, 1413 (1957)
117 W. Pauli (ed.): *Niels Bohr and the Development of Physics* (Pergamon, Oxford, New York 1955)
118 T.D. Lee, C.N. Yang: Phys. Rev. **104**, 254 (1956)
119 L.B. Okun: Yad. Fiz. **1**, 938 (1965);
and T.D. Lee, L. Wolfenstein: Phys. Rev. **B138**, 1490 (1965)
120 J. Bernstein, G. Feinberg, T.D. Lee: Phys. Rev. **B139**, 1650 (1965)
121 A. Salam: Contemp. Phys. **1**, 337 (1960)
122 R.P. Feynman, M. Gell-Mann: Phys. Rev. **109**, 193 (1958);
and R.E. Marshak, E.C. Sudarshan: Phys. Rev. **109**, 1860 (1958)
123 C.S. Wu, S.A. Moszkowski: *Beta Decay* (Wiley Interscience, New York 1966)
124 D. Tadic: Rep. Progr. Phys. **43**, 67 (1980)
125 N. Cabbibo: Phys. Rev. Lett. **10**, 531 (1963)
126 O.P. Sushkov, V.V. Flambaum: Usp. Fiz. Nauk **136**, 3 (1982)
127 I.S.Shapiro: Usp. Fiz. Nauk **95**, 647 (1968)
128 R.J. Blin-Styole: Phys. Rev. **118**, 1605 (1960); Phys. Rev. **120**, 181 (1961)
129 G.V. Danilyan: *Physics of Atomic Nucleus and Elementary Particles*, Proc. Conf. Nuclear Physics Research, Kharkov, Oct. 1982, Pt. 1 (TsNIIAtominform, Moscow 1983) pp. 205–211
130 M.T.Burgy, R.J. Epstein, V.E. Krohn, T.B. Novey, S. Raboy, G.R. Ringo, V.L. Telegdi: Phys. Rev. **107**, 1731 (1957)
131 B.G. Erozolimsky: Usp. Fiz. Nauk **116**, 145 (1975)
132 M.T. Burgy, V.E. Krohn, T.B. Novey, G.R. Ringo, V.L. Telegdi: Phys. Rev. **120**, 1829 (1960)
133 M.A. Clark, J.M. Robson: Can. J. Phys. **38**, 693 (1960); Can. J. Phys. **39**, 13 (1961)
134 V. Krohn, G. Ringo: Phys. Lett. **B55**, 175 (1975)

135 B.G. Erozolimsky, L.N. Bondarenko, Yu. A. Mostovoi, B.A. Obinyakov, V.A. Titov, V.P. Zakharova, A.I. Frank: Yad. Fiz. **12**, 323 (1970)
136 B.G. Erozolimsky, A.I. Frank, Yu.A. Mostovoi, S.S. Arzumanov, L.P. Vojzik: Yad. Fiz. **30**, 692 (1979)
137 B.G. Erozolimsky, L.N. Bondarenko, Yu.A. Mostovoi, B.A. Obinyakov, V.P. Fedunin, A.I. Frank: Pis'ma Zh. Eksp. Teor. Fiz. **13**, 356 (1971)
138 T.D. Lee, C.N. Yang: Phys. Rev. **105**, 1671 (1957);
and L.D. Landau: Zh. Eksp. Teor. Fiz. **32**, 407 (1957);
and A. Salam: Nuovo Cimento **5**, 299 (1957)
139 *Review of Particle Properties:* Phys. Lett. **B170**, 1 (1986)
140 V.A. Lubimov, E.G. Novikov, V.Z. Nozik, E.F. Tretyakov, V.S. Kosik: Phys. Lett. **B94**, 266 (1980); and Yad. Fiz. **32**, 301 (1980)
141 B.G. Erozolimsky, L.N. Bondarenko, Yu.A. Mostovoi, B.A. Obinyakov, V.P. Zakharova, V.A. Titov: Yad. Fiz. **8**, 176 (1968); and Yad. Fiz. **11**, 1049 (1970)
142 A.I. Frank: Usp. Fiz. Nauk **137**, 5 (1982)
143 B.G. Erozolimsky, Yu.A. Mostovoi, V.P. Fedunin, A.I. Frank, O.V. Khakhan: Pis'ma Zh. Eksp. Teor. Fiz. **20**, 745 (1974)
144 R.I. Steinberg, P. Liaud, B. Vignon, V.W. Hughes: Phys. Rev. Lett. **33**, 41 (1974); and Phys. Rev. **D13**, 2469 (1976)
145 B.G. Erozolimsky, Yu.A. Mostovoi, V.P. Fedunin, A.I. Frank, O.V. Khakhan: Yad. Fiz. **28**, 98 (1978)
146 Yu.G. Abov, P.A. Krupchitsky, Yu.A. Oratovsky: Phys. Lett. **12**, 25 (1964); and Yad. Fiz. **1**, 479 (1965)
147 D.H. Wilkinson: Phys. Rev. **109**, 1603 (1958)
148 R.J. Blin-Stoyle: *Fundamental Interactions and the Nucleus* (North-Holland, Amsterdam 1973)
149 Yu.G. Abov, P.A. Krupchitsky: Usp. Fiz. Nauk **118**, 141 (1976)
150 L.C. Biedenharn, M.E. Rose: Rev. Mod. Phys. **25**, 729 (1953)
151 *Alpha, Beta- and Gamma-Ray Spectroscopy*, ed. by K. Siegbahn (North-Holland, Amsterdam 1965)
152 G.A. Lobov: Izv. Akad. Nauk SSSR, Ser. Fiz. **32**, 886 (1968)
153 R. Haas, L. Leipuner, R. Adair: Phys. Rev. **116**, 1221 (1959)
154 Yu.G. Abov, O.N. Ermakov, P.A. Krupchitsky: Zh. Eksp. Teor. Fiz. **65**, 1738 (1973)
155 M. Forte, O. Saavedra: Preprint EUR, 3053e (Ispra 1966)
156 J. Eichler, P. Heine: Z. Phys. **227**, 352 (1969)
157 E. Warming, F. Stecher-Rasmussen, W. Ratynski, J. Kopecky: Phys. Lett. **B25**, 200 (1967)
158 E. Warming: Phys. Lett. **B29**, 564 (1969)
159 G.V. Danilyan, V.V. Novitsky, V.S. Pavlov, S.P. Borovlev, B.D. Vodennikov, V.P. Dronyaev: Pis'ma Zh. Eksp. Teor. Fiz. **24**, 380 (1976)
160 J.L. Alberi, R. Wilson, J.G. Schröder: Phys. Rev. Lett. **29**, 518 (1972)
161 H. Benkoula, J.F. Cavaignac, J.L. Charvet, D.H. Koang, B. Vignon, R. Wilson: Phys. Lett. **B71**, 287 (1977)
162 J.F. Cavaignac, B. Vignon, R. Wilson: Phys. Lett. **B67**, 148 (1977)
163 M. Avenier, J.F. Cavaignac, D.H. Koang, B. Vignon, R. Hart, R. Wilson: Phys. Lett. **B137**, 125 (1984)
164 A.N. Moskalev: Yad. Fiz. **9**, 163 (1969);
and B. Desplanques, J.F. Donahue, B.R. Holstein: Ann. Phys. N.Y. **124**, 449 (1980)
165 V.A. Vesna, E.A. Kolomensky, V.M. Lobashev, V.A. Nazarenko, A.N. Pirozhkov, L.M. Smotritsky, Yu.V. Sobolev, N.A. Titov: Pis'ma Zh. Eksp. Teor. Fiz. **36**, 169 (1982)
166 B.H.J. McKellar: Phys. Rev. **178**, 2160 (1969)
167 G.S. Danilov: Phys. Lett. **18**, 40 (1965)
168 V.A. Knyazkov. E.A. Kolomensky, V.M. Lobashev, V.A. Nazarenko, A.N. Pirozhkov, Yu.V. Sobolev, A.I. Shablij, E.V. Shulgina: Pis'ma Zh. Eksp. Teor. Fiz. **38**, 138 (1983)
169 L.B. Okun: Usp. Fiz. Nauk **89**, 603 (1966)
170 E.M. Henley, B.A. Jacobsohn: Phys. Rev. **113**, 225–233, 234–238 (1959)
171 S.P. Lloyd: Phys. Rev. **81**, 161 (1951)
172 P.A. Krupchitsky: Yad. Fiz. **3**, 974 (1966)
173 P.A. Krupchitsky, G.A. Lobov: At. Energy Rev. **7**, 91 (1969)

174 E.M. Henley, B.A. Jacobsohn: Phys. Rev. Lett. **16**, 706 (1966)
175 E.V. Tretyakov, G.V. Danilyan, V.S. Pavlov, G.I. Grishuk, V.E. Konyaev: Yad. Fiz. **7**, 7 (1968)
176 J. Kajfosz, J. Kopecky, J. Honzatko: Phys. Lett. **20**, 284 (1966); and Nucl. Phys. **A120**, 225 (1968)
177 P. Sharman, J.F. Cavaignac, J.-L. Charvet, W.D. Hamilton, P. Hungerford, B. Vignon: J. Phys. G: Nucl. Phys. **4**, 973 (1978)
178 J. Eichler: Nucl. Phys. **A120**, 535 (1968); and Nucl. Phys. **A127**, 693 (1969)
179 M.I. Bulgakov, G.W. Danilyan, A.D. Gulko, I.L. Karpikhin, P.A. Krupchitsky, V.V. Novitsky, Yu.A. Oratovsky, V.S. Pavlov, E.I. Tarkovsky, S.S. Trostin: Phys. Lett. **B42**, 351 (1972); and Yad. Fiz. **18**, 12 (1973)
180 G.W. Wang, A.J. Becker, L.M. Chirovsky, J.L. Groves, C.S. Wu: Phys. Rev. **C18**, 476 (1978)
181 O.C. Kistner: Phys. Rev. Lett. **19**, 872 (1967)
182 N.K. Cheung, H.E. Henrikson, F. Boehm: Phys. Rev. **C16**, 2381 (1977); and J.L. Gimlett, H.E. Henrikson, N.K. Cheung, F. Boehm: Phys. Rev. **C24**, 620 (1981); and Phys. Rev. **C25**, 1567 (1982)
183 T.D. Hannon, G.T. Trammel: Phys. Rev. Lett. **21**, 726 (1968)
184 B.T. Murdoch, C.E. Olsen, W.A. Steyert: J. Phys. G: Nucl. Phys. **3**, 1093 (1977)
185 F.L. Shapiro: Usp. Fiz. Nauk **65**, 133 (1958)
186 A.D. Gulko, S.S. Trostin, A. Khudoklin: Yad. Fiz. **6**, 657 (1967)
187 A.D. Gulko. S.S. Trostin, A. Khudoklin: Zh. Eksp. Teor. Fiz. **52**, 1504 (1967)
188 A. Winnacker, H. Ackermann, D. Dubbers, J. Mertens, P.V. Blanckenhagen: Z. Phys. **244**, 289 (1971)
189 J. Kranendonk: Physica **20**, 781 (1954)
190 D. Dubbers, K. Dörr, H. Ackermann, F. Fujara, H. Grupp, P. Heitjans, A. Körblein, H.-J. Stöckmann: Z. Phys. **A282**, 243 (1977)
191 M.T. Burgy, W.C. Davidon, T.B. Novey, G.J. Perlow, R. Ringo: Bull. Am. Phys. Soc. **2**, 206 (1957)
192 D.W. Connor: Phys. Rev. Lett. **3**, 429 (1959)
193 A.H. Wapstra, D.W. Connor: Nucl. Phys. **22**, 336 (1961)
194 Yu.G. Abov, O.N. Ermakov, A.D. Gulko, P.A. Krupchitsky, S.S. Trostin: Nucl. Phys. **34**, 505 (1961)
195 D.W. Connor, T. Tsang: Phys. Rev. **126**, 1506 (1962); and T. Tsang, D.W. Connor: Phys. Rev. **132**, 1141 (1963)
196 A.H. Wapstra, G.J. Nijgh, R. van Lieshout: *Nuclear Spectroscopy Tables* (North-Holland, Amsterdam 1959)
197 H. Rauch: Z. Phys. **197**, 373 (1966)
198 H. Ackermann, D. Dubbers, J. Mertens, A. Winnacker, P.V. Blanckenhagen: Phys. Lett. **B29**, 485 (1969); and Z. Phys. **228**, 329 (1969)
199 H. Lades: Z. Phys. **252**, 242 (1972)
200 H.-J. Stöckmann, H. Ackermann, D. Dubbers, M. Grupp, P. Heitjans: Z. Phys. **269**, 47 (1974); and H. Ackermann, D. Dubbers, M. Grupp, P. Heitjans, H.-J. Stöckmann: Phys. Lett. **B52**, 54 (1974)
201 A. Winnacker, H. Ackermann, D. Dubbers, M. Grupp, P. Heitjans, H.-J. Stöckmann: Nucl. Phys. **A261**, 261 (1976)
202 H. Grupp, K. Dörr, H.-J. Stöckmann, H. Ackermann, B. Bader, W. Buttler, P. Heitjans: Z. Phys. **B47**, 1 (1982); and W. Buttler, H.-J. Stöckmann, H. Ackermann, K. Dörr, F. Fujara, H. Grupp, P. Heitjans, G. Kiese, A. Körblein, D.A. Dubbers: Z. Phys. **B45**, 273 (1982)
203 M.I. Bulgakov, A.D. Gulko, Yu.A. Oratovsky, S.S. Trostin: Zh. Eksp. Teor. Fiz. **61**, 667 (1971); and M.I. Bulgakov, S.P. Borovlev, A.D. Gulko, F.S. Dzheparov, S.S. Trostin: Pis'ma Zh. Eksp. Teor. Fiz. **27**, 481 (1978)
204 H. Ackermann, D. Dubbers, H. Grupp, P. Heitjans, H.-J. Stöckmann: Phys. Lett. **A54**, 399 (1975)
205 Yu.G. Abov, M.I. Bulgakov, A.D. Gulko, F.S. Dzheparov, S.S. Trostin, S.P. Borovlev, V.M. Garochkin: Pis'ma Zh. Eksp. Teor. Fiz. **35**, 344 (1982)
206 G.A. Lobov, G.V. Danilyan: Izv. Akad. Nauk SSSR, Ser. Fiz. **41**, 1548 (1977)

207 V.A. Vesna, A.I. Egorov, E.A. Kolomensky, V.M. Lobashev, A.N. Pirozhkov, L.M. Smotritsky, N.A. Titov: Pis'ma Zh. Eksp. Teor. Fiz. **33**, 429 (1981)
208 N.V. Borovikova, V.A. Vesna, A.I. Egorov, K.A. Knyazkov, E.A. Kolomensky, V.M. Lobashev, A.N. Pirozkhov, L.A. Popeko, L.M. Smotritsky, N.A. Titov, A.I. Shablij: Pis'ma Zh. Eksp. Teor. Fiz. **30**, 527 (1979)
209 V.M. Lobashev, V.A. Nazarenko, L.F. Saenko, L.M. Smotritsky, G.I. Kharkevich: Pis'ma Zh. Eksp. Teor. Fiz. **3**, 268 (1966)
210 Yu. P. Emelin, O.N. Ermakov, I.L. Karpikhin, P.A. Krupchitsky, V.F. Perepelitsa, B.G. Peskov: Preprint ITEP 142 (Moscow 1983)
211 O.P. Sushkov, V.V. Flambaum: Yad. Fiz. **33**, 629 (1981)
212 M. Goldberger, K. Watson: *Collision Theory* (Wiley, New York 1964)
213 P.J.J. Kok, J.B.M. Haas, K. Abrahams, H. Postma, W.J. Huiskamp: Z. Phys. **A234**, 271 (1986)
214 O.N. Ermakov, I.L. Karpikhin, P.A. Krupchitsky, G.A. Lobov, V.F. Perepelitsa, F. Stecker-Rasmussen, P.J.J. Kok: *Neutron Physics*, Proc. 6th USSR Conf. Neutron Physics, Kiev, 1983 (TsNIIAtominform, Moscow 1984) pp. 403–407; and Yad. Fiz. **43**, 1359 (1986)
215 V.A. Vesna, A.I. Egorov, E.A. Kolomensky, A.F. Konyushkin, V.M. Lobashev, I.S. Okunev, B.G. Peskov, A.N. Pirozkhov, L.M. Smotritsky, N.A. Titov, E.V. Shulgina: Pis'ma Zh. Eksp. Teor. Fiz. **38**, 265 (1983)
216 A. Antonov, V.A. Vesna, Yu.M. Gledenov, V.M. Lobashev, I.S. Okunev, Yu.P. Popov, H. Rigol, L.M. Smotritsky: Pis'ma Zh. Eksp. Teor. Fiz. **40**, 209 (1984)
217 V.V. Vladimirsky, V.N. Andreev: Zh. Eskp. Teor. Fiz. **41**, 663 (1961)
218 A. Bohr: Proc. Int. Conf. *Peaceful Uses of Atomic Energy*, Vol. 2, Geneva, 1955, (United Nations, New York, 1956) pp. 151–154
219 A.D. Budnik, N.S. Rabotnov: Phys. Lett. **B46**, 155 (1973)
220 G.V. Danilyan, V.P. Dronyaev, B.D. Vodennikov, V.V. Novitsky, V.S. Pavlov, S.P. Borovlev: Preprint ITEP 4 (Moscow 1977)
221 G.V. Danilyan: Usp. Fiz. Nauk **131**, 329 (1980)
222 O. Bohr, B. Mottelson: *Nuclear Structure*, Vol. 2: *Nuclear Deformations* (Benjamin, New York 1974)
223 G.V. Danilyan, B.D. Vodennikov, V.P. Dronyaev, V.V. Novitsky, V.S. Pavlov, S.P. Borovlev: Pis'ma Zh. Eksp. Teor. Fiz. **26**, 197 (1977); and Yad. Fiz. **27**, 42 (1978)
224 B.D. Vodennikov, G.V. Danilyan, V.P. Dronyaev, V.V.Novitsky, V.S. Pavlov, S.P. Borovlev: Pis'ma Zh. Eksp. Teor. Fiz. **27**, 68 (1978)
225 V.A. Vesna, V.A. Knyazkov, E.A. Kolomensky, V.M. Lobashev, A.N. Pirozhkov, L.A. Popeko, L.M. Smotritsky, S.M. Solovyov, N.A. Titov: Pis'ma Zh. Eksp. Teor. Fiz. **31**, 704 (1980)
226 B.D. Vodennikov, G.V. Danilyan, V.P. Dronyaev, V.V. Novitsky, V.S. Pavlov, E.S. Rzhevsky: Preprint ITEP 137 (Moscow 1980)
227 A.Ya. Aleksandrovich, D.V. Nikolaev, T.K. Zvezdkina, G.A. Petrov, A.K. Petukhov, Yu.S. Pleva, S.M. Solovyov: Preprint LINP 797 (Leningrad 1982)
228 A.K. Petukhov, G.A. Petrov, S.I. Stepanov, D.V. Nikolaev, T.K. Zvezdkina, V.I. Petrova, V.A. Tyukavin: Pis'ma Zh. Eksp. Teor. Fiz. **30**, 470 (1979); and Pis'ma Zh. Eksp. Teor. Fiz. **32**, 324 (1980)
229 A.Ya. Aleksandrovich, T.K. Zvezdkina, D.V. Nikolaev, G.A. Petrov, A.K. Petukhov, Yu.S. Pleva: Preprint LINP 804 (Leningrad 1982); and Yad. Fiz. **39**, 805 (1984)
230 V.P. Alfimenkov, B.D. Vodennikov, G.V. Danilyan, V.P. Dronyaev, A.I. Ivanenko, Yu.D. Mareev, V.V. Novitsky, L.B. Pikelner, S.M. Solovyov: Preprint ITEP 49 (Moscow 1981)
231 G.V. Val'sky, T.K. Zvezdkina, D.V. Nikolaev, G.A. Petrov, Yu.S. Pleva, V.I. Petrova, V.A. Tyukavin: Yad. Fiz. **39**, 276 (1984)
232 V.A. Vesna, E.A. Kolomensky, A.F. Kornyushkin, V.M. Lobashev, I.S. Okunev, A.N. Pirozhkov, L.M. Smotritsky, S.M. Solovyov, N.A. Titov, E.V. Shulgina: Pis'ma Zh. Eksp. Teor. Fiz. **37**, 410 (1983)
233 A.G. Beda, L.N. Bondarenko, B.D. Vodennikov, G.V. Danilyan, V.P. Dronyaev, S.V. Zhukov, V.M. Kolobashkin, E.I. Korobkina, V.L. Kuznetsov, V.A. Kutsenko, Yu.A. Mostovoi, V.V. Novitsky, V.S. Pavlov, Yu. F. Pevchev, A.G. Sadchikov: Pis'ma Zh. Eksp. Teor. Fiz. **38**, 141 (1983)

234 V.N. Andreev, M.M. Danilov, O.N. Ermakov, V.G. Nedopekin, V.I. Rogov: Pis'ma Zh. Eksp. Teor. Fiz. **28**, 53 (1978); and Yad. Fiz. **30**, 306 (1979);
and V.V. Andreev, M.M. Danilov, Yu. D. Katarzhnov, V.D. Nedopekin, V.I. Rogov: Pis'ma Zh. Eksp. Teor. Fiz. **31**, 564 (1980)
235 J.S. Fraser: Phys. Rev. **88**, 536 (1952)
236 O.P. Sushkov, V.V. Flambaum: Phys. Lett. **B94**, 277 (1980); and Yad. Fiz. **33**, 59 (1981)
237 F.C. Michel: Phys. Rev. **B133**, 329 (1964)
238 L. Stodolsky: Phys. Lett. **B50**, 352 (1974)
239 M. Forte: ILL Research Proposal 03-03-002 (Grenoble 1976)
240 M. Forte, B.R. Heckel, N.F. Ramsey, K. Green, G.L. Greene, J. Byrne, J.M. Pendlebury: Phys. Rev. Lett. **45**, 2088 (1980)
241 M. Forte: *Inst. Phys. Conf. Ser.* 42, Chap. 2 (Inst. Phys. London 1978) pp. 86–100
242 L. Stodolsky: Phys. Lett. **B96**, 127 (1980)
243 O.P. Sushkov, V.V. Flambaum: Pis'ma Zh. Eksp. Teor. Fiz. **32**, 377 (1980)
244 E.A. Kolomensky, V.M. Lobashev, A.N. Pirozhkov, L.M. Smotritsky, N.A. Titov, V.A. Vesna: Phys. Lett. **B107**, 272 (1981)
245 V.A. Vesna, E.A. Kolomensky, V.M. Lobashev, A.N. Pirozhkov, L.M. Smotritsky, N.A. Titov: Pis'ma Zh. Eksp. Teor. Fiz. **35**, 351 (1982)
246 V.P. Alfimenkov, S.B. Borzakov, Vo Van Thuan, Yu. D. Mareev, L.B. Pikelner, D. Rubin, A.S. Khrikin, E.I. Sharapov: Pis'ma Zh. Eksp. Teor. Fiz. **34**, 308 (1981); **35**, 42 (1982)
247 V.P. Alfimenkov, S.B. Borzakov, Vo Van Thuan, Yu.D. Mareev, L.B. Pikelner, A.S. Khrikin, E.I. Sharapov: Nucl. Phys. **A398**, 93 (1983)
248 V.A. Karmanov, G.A. Lobov: Pis'ma Zh. Eksp. Teor. Fiz. **10**, 332 (1969)
249 V.E. Bunakov, V.P. Gudkov: Z. Phys. **A303**, 285 (1981);
and G.A. Lobov: Yad. Fiz. **35**, 1408 (1982); and Preprint ITEP 20 (Moscow 1982)
250 V.P. Alfimenkov: Usp. Fiz. Nauk **144**, 361 (1984)
251 L. Stodolsky: Nucl. Phys. **B197**, 213 (1982)
252 D.F. Zaretsky, V.K. Sirotkin: Yad. Fiz. **37**, 607 (1983); and Yad. Fiz. **39**, 585 (1984); and Yad. Fiz. **42**, 885 (1985)
253 B. Heckel, N.F. Ramsey, K. Green, G.L. Greene, R. Gähler, O. Schaerf, M. Forte, W. Dress, P.D. Miller, R. Golub, J. Byrne, J.M. Pendlebury: Phys. Lett. **B119**, 298 (1982)
254 Yu.G. Abov, O.N. Ermakov, I.L. Karpikhin, P.A. Krupchitsky, Yu. E. Kuznetsov, V.F. Perepelitsa, V.I. Petrushin: Preprint ITEP 181 (Moscow 1983); and Yad. Fiz. **40**, 1585 (1984)
255 B. Heckel, M. Forte, O. Schaerf, K. Green, G.L. Greene, N.F. Ramsey, J. Byrne, J.M. Pendlebury: Phys. Rev. **C29**, 2389 (1984)
256 V.P. Alfimenkov, S.B. Borzakov, Vo Van Thuan, Yu.D. Mareev, L.B. Pikelner, I.M. Frank, A.S. Khrikin, E.I. Sharapov: Pis'ma Zh. Eksp. Teor. Fiz. **39**, 346 (1984)

Subject Index

Adiabaticity condition 2, 7
Anisotropy of angular distribution of fission fragments 106
–––––––– as a function of mass 109
–––––––– as a function of energy 109
– – alpha emission 98
– – beta emission 89
– – gamma emission 44
– – – –, P odd 71
Asymmetry coefficient in alpha decay 99
– – – beta decay 57
– – – gamma decay 72
–, instrumental 77
– of radiative-capture cross sections 119, 125
– – total-interaction cross sections 119, 125
– – transmission of longitudinally polarized
– –   neutrons 119

Beam polarization 3
– –, measurement by double reflection 4
– –, – – shim technique 5

Cabibbo angle 60
– model 60
Centrifugal barrier 99
Charge conjugation 55
– parity 56
– –, violation of 57
Charged currents 60
Circular polarization 50
Coefficient of asymmetry of fission fragment emission 106, 110
– – left-right asymmetry 99, 103, 113
Coherent amplitude of nuclear scattering 118
– interference of spin states 49
– length of magnetic scattering 16
– – – nuclear scattering 15
Collimators, neutron beam 6
Combined (CP) parity 56
– – –, violation of 23, 36, 57
Conservation laws, quantum mechanical 55
Correlation, antineutrino-neutron spin 64, 65
– coefficients 67, 70
–, electron-neutron spin 64, 65
–, neutron spin-electron-antineutrino 65
– of cascade gamma quanta 83

Correlations, P-even 78, 99, 112
Coupling constants of the weak interaction 60
Critical glancing angle 15

Decay of polarized neutrons 64
Depolarization of neutron beam 5
Depth of neutron penetration 15
Detector, neutron 7
Distribution, angular, of gamma rays 44, 72

Electric dipole moment of neutron 28
– – – – –, upper bound of 35
Electroweak interaction 58

Helicity of antineutrino 67
– – gamma quanta 50
– – neutrons 118

Integral method of recording 7, 80, 98
Interference of s- and p-wave resonances 99, 112
Invariance under time reversal 56, 68, 83
Inversion of coordinates 55

Lagrangian, four-fermion 59
Larmor precession of spins 1, 24
Leading magnetic field 4
Luders-Pauli theorem (CPT theorem) 56

Magnetic moment of the neutron, magnitude of 27
– – – – –, measurement of 25
– – – – –, sign of 27
– moments, nuclear 93, 94
– resonance, multispin 97
Mixed compound states, model of 120
Mixing of gamma transition 73
– – – –, parameter of 73, 83
– – spin channels 46
– – – –, parameter of 46
– – states with opposite parities, 62
– – – – – –, coefficient of 62

Neutral currents 60
Neutron guides 14

137

– spin, behavior in magnetic field 1
Nonadiabatic spin reversal 2
Nonregular electromagnetic transition 72
Nuclear magnetic resonance of polarized
    nuclei 93
Nucleon-nucleon weak interaction,
    enhancement mechanism of 61
– – – –, isotopic structure of 60, 81
– – – –, one-particle potential of 61
– – – –, relative value of 61
– – – –, structure of 60

Phase difference of matrix elements 83, 87
Polarization cross section 38
– efficiency 3
– function 50
– of compound nuclei 89
– – neutron beam 3
– – nuclei 37
– ratio 4
Polarized deuterium target 40
– proton target 40
Precession of neutrons, nuclear 42
Probability of spin flip 4

Quadrupole nuclear moments 93, 94
Quark multiplet 60

Reflection of neutrons, total 15
Refractive index of a medium 15, 17
Regular electromagnetic transition 72
Reversal of spin, nonadiabatic 3
Rotation of spin, adiabatic 2, 7
– – spins of transversely polarized
        neutrons 121

Selection rules in nuclear beta decay 59
Shim ratio 5
Spatial parity 56
– –, violation of, in beta decay 57
– –, – –, in gamma decay 71
– –, – –, in neutron decay 65
– –, – –, in neutron optics 117
Spectrum of gamma quanta, integral 81
Spin channels 38, 44, 89
– flippers 8
– –, adiabatic 12
– –, nonadiabatic 9
– –, resonance 8
Supermirrors, polarizing 21

Target transmission 38
Time reversal 55
Transition, nonregular 72
–, regular 72

Variants of weak interaction 59

Weak interaction 57, 58
– –, Fermi coupling constant of 60
– –, nucleon-nucleon, Hamiltonian of 61
– –, universality of 58, 60
– –, variants of 59
– –, $V - A$ variant of 60, 67
Weinberg angle 61